Business Calculus

HARPERCOLLINS COLLEGE OUTLINE

Business Calculus

Ron Smith, Ph.D.
Edison Community College

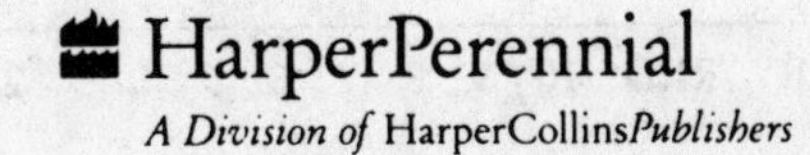

HarperPerennial
A Division of HarperCollins*Publishers*

 For information address HarperCollins Publishers, Inc., 10 East 53rd Street, New York, NY 10022.

An American BookWorks Corporation Production

Project Manager: Jonathon E. Brodman
Editor: Gloria Langer

Library of Congress Cataloging-in-Publication Data

Smith, Ron 1949–
Business calculus / Ron Smith.
p. cm. — (HarperCollins college outline)
Includes index.
ISBN: 0-06-467136-4
1. Calculus. I. Title. II. Series.
QA303.S6544 1993
515—dc20 91-58276

93 94 95 96 97 ABW/RRD 10 9 8 7 6 5 4 3 2 1

Contents

Preface

Business Calculus is designed to be a companion text to that which is used in any business calculus course. This book contains the topics commonly found in the courses on business calculus taught at colleges and universities.

Each topic is addressed with the basic concepts in mind. The text examples are extensively annotated at each point along with step-by-step solutions. Many often-asked questions are answered during the solution process. Exercises that mirror the examples are included at the end of each chapter to check for understanding, so that the student can return to the primary text for the course.

I hope that this text provides insight by which a student can build a strong knowledge of calculus.

Ron Smith

Business Calculus

1

Algebra for Calculus

1.1 FACTORING

Introduction

Many students who have difficulty with calculus find that their difficulty lies not with the calculus concepts, but with an unfamiliarity with algebraic techniques. This chapter will emphasize the algebra needed for calculus.

Prior to reading this section, please make sure that you are comfortable with (a) removing common factors, (b) factoring by pairing, and (c) trinomial factoring. These topics are covered in *Elementary Algebra* or *Intermediate Algebra*—other books in this series.

Factoring concepts covered here will be:

a) Factoring the sum of two cubes and the difference of two cubes
b) Using radicals in factoring
c) Removing common factors from complicated binomials

The Sum and Difference of Two Cubes

The general formulas for the sum of two cubes and the difference of two cubes are as shown below. The letters a and b represent any algebraic term.

1.1 Sum of Two Cubes

$$a^3 + b^3 = (a+b)(a^2 - ab + b^2)$$

1.2 Difference of Two Cubes

$$a^3 - b^3 = (a-b)(a^2 + ab + b^2)$$

Please note that $a^3 + b^3$ does *not* equal $(a+b)^3$ and that $a^3 - b^3$ does *not* equal $(a-b)^3$. The factorization of the sum of two cubes and the difference of two cubes is demonstrated in the following example.

EXAMPLE 1.1

Use the formulas for the sum and difference of two cubes to factor the following.

a) x^3+8y^3
b) $27x^3-1$
c) $8a^3+b^6$
d) c^3d^3-64

SOLUTION 1.1

a) x^3+8y^3 — Copy the original binomial.

$x^3+2^3y^3$ — Rewrite 8 as 2^3.

$(x)^3+(2y)^3$ — Referring to the general formula, 1.1, let $a = x$ and $b = 2y$.

$a^3+b^3 = (a+b)(a^2-ab+b^2)$ — Place the new terms in the format of the general formula.

$$(x)^3+(2y)^3 = (x+2y)((x)^2-(x)(2y)+(2y)^2)$$

$$x^3+8y^3 = (x+2y)(x^2-2xy+4y^2)$$

Multiply within terms and simplify.

b) $27x^3-1$ — Copy the original binomial.

$3^3x^3-1^3$ — Rewrite 27 as 3^3, and 1 as 1^3.

$(3x)^3-(1)^3$ — Referring to the general formula, 1.2, let $a = 3x$ and $b = 1$.

$a^3-b^3 = (a-b)(a^2+ab+b^2)$ — Place the new terms in the format of the general formula.

$$(3x)^3-(1)^3 = (3x-1)((3x)^2+(3x)(1)+1^2)$$

$$27x^3-1 = (3x-1)(9x^2+3x+1)$$

Multiply within terms and simplify.

c) $8a^3 + b^6$ — Copy the original binomial.

$2^3a^3 + (b^2)^3$ — Rewrite 8 as 2^3 and b^6 as $(b^2)^3$.

Note that $b^6 = b^{2 \cdot 3} = (b^2)^3$.

$(2a)^3 + (b^2)^3$ — Use the general formula, 1.1, letting $a = 2a$ and $b = b^2$.

$a^3 + b^3 = (a+b)(a^2 - ab + b^2)$ — Place the new terms in the format of the general formula.

$$(2a)^3 + (b^2)^3 = (2a + b^2)((2a)^2 - (2a)(b^2) + (b^2)^2)$$

$$8a^3 + b^6 = (2a + b^2)(4a^2 - 2ab^2 + b^4)$$

Multiply within terms and simplify.

d) $c^3d^3 - 64$ — Copy the original binomial.

$c^3d^3 - 4^3$ — Rewrite 64 as 4^3.

$(cd)^3 - (4)^3$ — Referring to the general formula, 1.2, let $a = cd$ and $b = 4$.

$a^3 - b^3 = (a-b)(a^2 + ab + b^2)$ — Place the new terms in the format of the general formula.

$$(cd)^3 - (4)^3 = (cd - 4)((cd)^2 + (cd)(4) + (4)^2)$$

$$c^3d^3 - 64 = (cd - 4)(c^2d^2 + 4cd + 16)$$

Multiply within terms and simplify.

Factoring Using Radicals

A calculus course is the first time radical expressions are used extensively in factoring. A radical sign is, as you may recall, the symbol $\sqrt{}$.

Usually, factoring with radical signs involves the use of the factoring technique for the **difference of two squares.**

By using radicals, every binomial with terms separated by a minus sign can be factored using the formula for the **difference of two squares.**

Note the following two formulas:

1.3 The Difference of Two Squares

$$a^2 - b^2 = (a+b)\ (a-b)$$

1.4 Using the Square Root Radical

$$a = \sqrt{a} \cdot \sqrt{a} = (\sqrt{a})^2$$

Formulas 1.3 and 1.4 are combined to form the basis for factoring using radical signs.

1.5 Factoring Using Radicals

$$a - b = (\sqrt{a})^2 - (\sqrt{b})^2 = (\sqrt{a}+\sqrt{b})\ (\sqrt{a}-\sqrt{b})$$

EXAMPLE 1.2

Factor the following binomials using Formula 1.5.

a) $x^2 - 3$

b) $y - 16$

c) $x - 2$

SOLUTION 1.2

a) $x^2 - 3$ — Copy the given binomial.

$(x)^2 - (\sqrt{3})^2$ — Rewrite x^2 as $(x)^2$ and use Formula 1.4 to convert 3 to $(\sqrt{3})^2$.

$(x+\sqrt{3})\ (x-\sqrt{3})$ — Using the difference of two squares process, 1.3, factor the binomial.

$x^2 - 3 = (x+\sqrt{3})\ (x-\sqrt{3})$ — The final answer is based on Formula 1.5.

b) $y - 16$ — Copy the given binomial.

$(\sqrt{y})^2 - 4^2$ — Rewrite 16 as 4^2 and use Formula 1.4 to convert y to $(\sqrt{y})^2$.

$(\sqrt{y}+4)\ (\sqrt{y}-4)$ — Using the difference of two squares process, 1.3, factor the binomial.

$y - 16 = (\sqrt{y} + 4)(\sqrt{y} - 4)$	The final answer is based on Formula 1.5.
c) $x - 2$	Copy the given binomial.
$(\sqrt{x})^2 - (\sqrt{2})^2$	Use Formula 1.4 to convert x to $(\sqrt{x})^2$ and 2 to $(\sqrt{2})^2$.
$(\sqrt{x} + \sqrt{2})(\sqrt{x} - \sqrt{2})$	Using the difference of two squares process, 1.3, factor the binomial.
$x - 2 = (\sqrt{x} + \sqrt{2})(\sqrt{x} - \sqrt{2})$	The final answer is based on Formula 1.5.

We will use Formula 1.5 in the chapter on limits (chapter 2).

Factoring Complicated Binomials

In subsequent chapters we will use a process called the Product Rule for derivatives. This process results in large binomials that can often be factored. The factorization of these binomials will be completed using the procedure outlined in the following exercise. This procedure is similar to removing a common factor from a polynomial.

> **1.6 Removing a Common Factor**
>
> $$ca + cb = c(a + b)$$

EXAMPLE 1.3

Factor the following expressions by removing common factors and simplifying.

a) $8(x-3)^4(2x-1)^2 + 2(x-3)^3(2x-1)^3$

b) $3(x-1)^2(3x-2)^7 + 7(x-1)^3(3x-2)^6$

c) $12(x+6)^5 + 4(x-2)(x+6)^4$

SOLUTION 1.3

a) $\underbrace{8(x-3)^4(2x-1)^2}_{\text{1 term}} + \underbrace{2(x-3)^3(2x-1)^3}_{\text{1 term}}$

Copy the large binomial.

$2[4(x-3)^4(2x-1)^2 + (x-3)^3(2x-1)^3]$

Remove a common factor of 2 from each term.

$2(x-3)^3[4(x-3)^1(2x-1)^2+(2x-1)^3]$

Remove three common factors of $(x-3)$ from each term (three factors are the largest number of factors found in each term).

$2(x-3)^3(2x-1)^2[4(x-3)+(2x-1)]$

Remove two common factors of $(2x-1)$ from each term (two factors are the largest number of factor present in both terms).

$2(x-3)^3(2x-1)^2[4x-12+2x-1]$

Distribute 4.

$2(x-3)^3(2x-1)^2[6x-13]$

Combine like terms.

$8(x-3)^4(2x-1)^2+2(x-3)^3(2x-1)^3$

$=2(x-3)^3(2x-1)^2(6x-13)$

b) $\underbrace{3(x-1)^2(3x-2)^7}_{\text{1 term}}+\underbrace{7(x-1)^3(3x-2)^6}_{\text{1 term}}$

Copy the large binomial.

$(x-1)^2[3(3x-2)^7+7(x-1)^1(3x-2)^6]$

Remove two common factors of $(x-1)$ from each term (two factors are the largest number of factors present in each term).

$(x-1)^2(3x-2)^6[3(3x-2)^1+7(x-1)^1]$

Remove six common factors of $(3x-2)$ from each term (six factors are the largest number of factors present in each term).

$(x-1)^2(3x-2)^6[9x-6+7x-7]$ Distribute 3 and 7.

$(x-1)^2(3x-2)^6[16x-13]$ Combine like terms.

$$3(x-1)^2(3x-2)^7+7(x-1)^3(3x-2)^6$$
$$=(x-1)^2(3x-2)^6(16x-13)$$

c) $\underbrace{12(x+6)^5}_{\text{1 term}} + \underbrace{4(x-2)(x+6)^4}_{\text{1 term}}$	Copy the large binomial.
$4[3(x+6)^5+(x-2)(x+6)^4]$	Remove the common factor of 4 from each term.
$4(x+6)^4[3(x+6)+(x-2)]$	Remove four common factors of $(x + 6)$ from each term (four factors are the largest number of factors present in each term).
$4(x+6)^4[3x+18+x-2]$	Distribute 3.
$4(x+6)^4[4x+16]$	Combine like terms.
$16(x+6)^4(x+4)$	Remove a factor of 4 and move it to the left of the factors.

$$12(x+6)^5+4(x-2)(x+6)^4$$
$$=16(x+6)^4(x+4)$$

1.2 RATIONAL EXPONENTS

Converting Radical Expressions Before some calculus operations can be completed, all algebraic expressions in radical form must be converted to rational (fractional) exponents. The rule that describes this conversion is:

> **1.7** $\sqrt[b]{x^a} = x^{a/b}$
> where a and b are positive whole numbers $[b \geq 2]$

Note: b is called the index of the radical.

EXAMPLE 1.4

Convert the following radical expressions to terms with rational exponents.

a) $\sqrt[4]{x^7}$

b) $\sqrt[6]{x^4}$

c) $\sqrt{x}$

SOLUTION 1.4

a) $\sqrt[4]{x^7}$ — Copy the expression.

$= x^{7/4}$ — Write the 7 from within the radical sign in the numerator of the exponent and the index, 4, in the denominator.

b) $\sqrt[6]{x^4}$ — Copy the expression.

$= x^{4/6}$ — The number 4 from within the radical sign is the numerator of the exponent. The index, 6, is the denominator.

$= x^{2/3}$ — Reduce the fraction, if possible.

c) $\sqrt{x}$ — Copy the expression.

$\sqrt[2]{x^1}$ — Express the understood 1 exponent of the x within the radical sign. Express the understood index of 2.

$= x^{1/2}$ — The 1 from within the radical sign is the numerator of the exponent and the index 2 is the denominator.

Adding Rational Exponents

The addition of rational exponents occurs when expressions containing the same base—but varying exponents—are multiplied. The addition of rational exponents can be stated as follows in Rule 1.8.

1.8 $x^{a/b} \cdot x^{c/d} = x^{ad/bd} \cdot x^{cb/bd} = x^{(ad+cb)/bd}$
where a, b, c, and d are integers and $b, d \neq 0$.

EXAMPLE 1.5

Compute the following products involving the multiplication of expression with rational exponents.

a) $(x^{2/3})(x^{1/5})$
b) $(x^{1/2})(x^{3})$
c) $(x^{1/3})(x^{1/2}+x^{3/2})$
d) $(x^{3/4}+x^{1/4})(x-x^{2})$

SOLUTION 1.5

a) $(x^{2/3})(x^{1/5})$ — Copy original factors.

$= x^{10/15} \cdot x^{3/15}$ — Using Rule 1.8, convert the two rational exponents so that they have the same denominator.

$= x^{(10+3)/15} = x^{13/15}$

b) $(x^{1/2})(x^{3})$ — Copy original factors.

$= x^{1/2} \cdot x^{3/1}$ — Rewrite the exponent as $\frac{3}{1}$ (a fraction).

$x^{1/2} \cdot x^{3/1} = x^{1/2} \cdot x^{6/2}$ — Using Rule 1.8, the two rational exponents are converted so that they have the same denominator.

$x^{1/2} \cdot x^{6/2} = x^{(1+6)/2} = x^{7/2}$ — Add the exponent numerators.

c) $(x^{1/3})(x^{1/2}+x^{3/2})$ — Copy the two expressions.

$x^{1/3} \cdot x^{1/2} + x^{1/3} \cdot x^{3/2}$ — Distribute $x^{1/3}$.

$x^{2/6} \cdot x^{3/6} + x^{2/6} \cdot x^{9/6}$ — Use Rule 1.8 to convert the exponent denominators to the same numbers.

$x^{(2+3)/6} + x^{(2+9)/6} = x^{5/6} + x^{11/6}$ — Add the numerators of the exponents in each term.

d) $(x^{3/4}+x^{1/4})(x-x^2)$ — Copy the two binomials.

$(x^{3/4}+x^{1/4})(x-x^2) = (x^{3/4}+x^{1/4})(x^{1/1}-x^{2/1})$ — Express the whole numbers as fractional exponents.

$= x^{3/4}\cdot x^{1/1} - x^{3/4}\cdot x^{2/1} + x^{1/4}\cdot x^{1/1} - x^{1/4}\cdot x^{2/1}$ — Multiply each part of one algebraic expression by every part of the other expression.

$= x^{3/4}x^{4/4} - x^{3/4}x^{8/4} + x^{1/4}x^{4/4} - x^{1/4}x^{8/4}$ — Use Rule 1.8 to convert the denominators of each rational exponent to the same number.

$= x^{(3+4)/4} - x^{(3+8)/4} + x^{(1+4)/4} - x^{(1+8)/4}$ — Add the numerators of the exponents.

$= x^{7/4} - x^{11/4} + x^{5/4} - x^{9/4}$

Subtracting Rational Exponents

The subtraction of rational exponents often occurs when one algebraic expression is divided by another. We will limit our discussion to the case in which the divisor is one term.

Three rules that assist in the subtraction of rational exponents are shown here:

1.9 $\dfrac{1}{x^{a/b}} = x^{-a/b}$

1.10 $x^{c/d}\cdot x^{-a/b} = x^{bc/bd}\cdot x^{-ad/bd} = x^{(bc-ad)/bd}$

1.11 $\dfrac{a+b+c+\ldots}{d} = \dfrac{a}{d}+\dfrac{b}{d}+\dfrac{c}{d}+\ldots$

Here, a, b, c, and d are real numbers and $b, d \neq 0$.

EXAMPLE 1.6

Simplify the following algebraic expressions, using subtraction of rational exponents.

a) $\dfrac{x^3}{x^5}$

b) $\dfrac{x^{2/3}}{x^{1/4}}$

c) $\dfrac{x^2 - x + 6}{x^{1/2}}$

d) $\dfrac{x^3 + x}{\sqrt[3]{x}}$

SOLUTION 1.6

a) $\dfrac{x^3}{x^5} = \dfrac{x^3}{1} \cdot \dfrac{1}{x^5}$ — Copy the original problem and rewrite as the product of two fractions.

$x^3 \cdot x^{-5}$ — Use Rule 1.9 to convert the factor, $1/x^5$, to one with a negative exponent.

$x^{3-5} = x^{-2}$ — Combine the two exponents through subtraction to obtain the final answer.

b) $\dfrac{x^{2/3}}{x^{1/4}} = \dfrac{x^{2/3}}{1} \cdot \dfrac{1}{x^{1/4}}$ — Copy the original problem and rewrite it as a product of two fractions.

$x^{2/3} \cdot x^{-1/4}$ — Use Rule 1.9 to convert the factor, $1/x^{1/4}$, to one with a negative exponent.

$x^{8/12} \cdot x^{-3/12}$ — Use Rule 1.10 to convert each exponent to a fraction with a common denominator.

$x^{(8-3)/12} = x^{5/12}$ — Subtract numerators of each exponent to obtain the final answer.

c) $\dfrac{x^2 - x + 6}{x^{1/2}}$

Use Rule 1.12 to separate the single fraction into a series of three fractions.

$$= \frac{x^2}{x^{1/2}} - \frac{x}{x^{1/2}} + \frac{6}{x^{1/2}}$$

$$= \frac{x^2}{1} \cdot \frac{1}{x^{1/2}} - \frac{x}{1} \cdot \frac{1}{x^{1/2}} + \frac{6}{1} \cdot \frac{1}{x^{1/2}}$$

Each fraction is separated into two factors. Note that $x = x^1$.

$$= x^2 \cdot x^{-1/2} - x^1 \cdot x^{-1/2} + 6 \cdot x^{-1/2}$$

Convert the exponent of each product into fractions with a common denominator. The right-most product needs no conversion.

$$= x^{(4-1)/2} - x^{(2-1)/2} + 6x^{-1/2}$$

$$= x^{3/2} - x^{1/2} + 6x^{-1/2}$$

Subtract the exponent numerators to complete the process.

d) $\dfrac{x^3 + x}{\sqrt[3]{x}} = \dfrac{x^3 + x^1}{x^{1/3}}$

Copy the problem and rewrite the denominator according to Rule 1.7. Note that $x = x^1$.

$$= \frac{x^3}{x^{1/3}} + \frac{x^1}{x^{1/3}}$$

Use Rule 1.12 to separate the single fraction into a series of two fractions.

$$= x^3 \cdot \frac{1}{x^{1/3}} + x^1 \cdot \frac{1}{x^{1/3}}$$

Each fraction is separated into two factors.

$$= x^3 \cdot x^{-1/3} + x^1 \cdot x^{-1/3}$$

Use Rule 1.9 to convert to negative exponents.

$$= x^{9/3} \cdot x^{-1/3} + x^{3/3} \cdot x^{-1/3}$$

Convert the exponents of each product into fractions with a common denominator.

$$= x^{(9-1)/3} + x^{(3-1)/3}$$

$$= x^{8/3} + x^{2/3}$$

Combine the numerators of the rational exponents and subtract.

1.3 FUNCTIONS

Functions and Function Notation

A **function** is a rule, process, or method that describes a correspondence between one set of elements (the **domain**) and a second set of elements (the **range**).

For each element in the domain there exists **one and only one** element in the range.

Through convention, a function is typically defined in terms of a variable, for example, x. When another variable, y, is set equal to this function, we say that y is a function of x, or $y = f(x)$. This expression is read "y is a function of x." $f(x)$ **does not** mean f times x.

When a member of the domain is placed into the function, it produces an element from the range. The association of these two elements gives rise to the term **ordered pair**.

The $f(x)$ notation is set equal to an expression in x. This expression then serves as the function of x. Replacement values for x, such as numbers, letters, or other expressions, are then acted upon by this function.

Each replacement for x constitutes an element in the domain, while the outcome from the function is an element in the range. This range element is usually labelled y **or** $f(x)$.

Every element in the domain is paired with an element in the range. This pairing is called an ordered pair. The order is specified as the x-value first and the y-value second. For example, the function defined by

$$f(x) = 2x+3$$

has an ordered pair of $(1, 5)$ because the substitution of the domain element, 1, in the function produces a range element of 5.

$$\begin{aligned} f(1) &= 2(1)+3 \\ &= 2+3 \\ &= 5 \end{aligned}$$

In some texts, the ordered pair is stated in the general form as (x, y). However, due to some of the formula derivations that are found in calculus textbooks, it is more useful to think of an ordered pair as $(x, f(x))$. In subsequent sections, we will use the $(x, f(x))$ concept of ordered pairs to help explain the formulation of the derivative.

Evaluating a Function

As mentioned earlier, a function acts upon an element in the domain. The outcome of this action is a member of the range and the evaluation of the function. When the function is defined in terms of the variable x, it uses x as a "placeholder" for any substitutions that may occur. For example, the function

$$f(x) = 3x^2 - 2x + 7$$

shows that we can substitute any element of the domain into the $3x^2$ term and the $-2x$ term. Since the term, +7, does not contain the variable, x, no substitution takes place there.

The evaluation of this particular function at –2 gives a range element of 23.

EXAMPLE 1.7

Evaluate each function at the given domain element. Write the ordered pair for each problem.

Function	Domain Element
a) $f(x) = 2x - 5$	3
b) $f(x) = 5x^2 - 2x + 1$	h
c) $f(x) = \sqrt{x} - 2$	9
d) $f(x) = x^3 - 2x + 1$	–2
e) $f(x) = 8x^2 + 3x - 2\sqrt{x}$	t
f) $f(x) = 5$	9
g) $f(x) = 5$	c
h) $f(x) = 3x^2 + 3x + 2$	$-x$
i) $f(x) = x^3 - 1$	$-c$

SOLUTION 1.7

a) $f(x) = 2x - 5$ — Copy the function to be evaluated at 3.

$f(3) = 2 \cdot 3 - 5$ — Replace every letter x with 3.

$f(3) = 6 - 5 = 1$ — Complete the arithmetic.

$f(3) = 1$	1 is the range element.
$(3, 1)$	Ordered pair
b) $f(x) = 5x^2 - 2x + 1$	Copy the function to be evaluated at *h*.
$f(h) = 5h^2 - 2h + 1$	Replace every letter *x* with the letter *h*.
$f(h) = 5h^2 - 2h + 1$	There is no arithmetic to complete in this problem.
	$5h^2 - 2h + 1$ is the element associated with range of $f(x)$.
$(h, 5h^2 - 2h + 1)$	Ordered pair
c) $f(x) = \sqrt{x} - 2$	Copy the function to be evaluated at 9.
$f(9) = \sqrt{9} - 2$	Replace every letter *x* with 9.
$f(9) = \sqrt{9} - 2 = 3 - 2 = 1$	Complete the arithmetic.
	1 is the range element.
$(9, 1)$	Ordered pair
d) $f(x) = x^3 - 2x + 1$	Copy the function to be evaluated at –2.
$f(-2) = (-2)^3 - 2(-2) + 1$	Replace every letter *x* with –2.
$f(-2) = (-2)^3 - 2(-2) + 1$ $= -8 + 4 + 1 = -4 + 1 = -3$	Complete the arithmetic.
	–3 is the range element.
$(-2, -3)$	Ordered pair
e) $f(x) = 8x^2 + 3x - 2\sqrt{x}$	Copy the function to be evaluated at *t*.
$f(t) = 8t^2 + 3t - 2\sqrt{t}$	Replace every letter *x* with the letter *t*.
	$8t^2 + 3t - 2\sqrt{t}$ is the range element.
$(t, 8t^2 + 3t - 2\sqrt{t})$	Ordered pair

f) $f(x) = 5$	Copy the function to be evaluated at 9.
$f(9) = 5$	Replace every letter x with the number 9.
	Since there is no x on the right side of the equation, the range element is 5.
$(9, 5)$	Ordered pair
g) $f(x) = 5$	Copy the function to be evaluated at c.
$f(c) = 5$	Replace every letter x with the letter c.
$f(c) = 5$	Note that the result is the same as when this function was evaluated at 9.
$(c, 5)$	Ordered pair
h) $f(x) = 3x^2 + 3x + 2$	Copy the function to be evaluated.
$f(-x) = 3(-x)^2 + 3(-x) + 2$	Replace every letter x with the symbol $-x$.
$= 3x^2 - 3x + 2$	The expression $3x^2 - 3x + 2$ is in the range of this function.
$(-x, 3x^2 - 3x + 2)$	Ordered pair
i) $f(x) = x^3 - 1$	Copy the function to be evaluated.
$f(-c) = (-c)^3 - 1$	Replace the letter x with $-c$.
$f(-c) = (-c)^3 - 1$	Complete the algebra.
$= (-c)(-c)(-c) - 1$	$-c^3 - 1$ is in the range of
$= -c^3 - 1$	the function.
$(-c, -c^3 - 1)$	Ordered pair

The Difference Quotient

The process of developing some of the basic rules of calculus uses a formula called the difference quotient.

1.12 The Difference Quotient

$$\frac{f(x+h)-f(x)}{h}$$

Using the difference quotient involves the evaluation of functions as shown in the previous section. The evaluated function is then substituted into the formula to obtain a solution.

Exercises involving the difference quotient require two things before the exercise can begin: the difference quotient formula and any given function formula.

The procedure used to complete an exercise with the difference quotient involves three steps:

1. Evaluate the given $f(x)$ formula at $f(x+h)$. This involves substituting the binomial $x+h$ in place of the letter x.
2. Replace the symbols $f(x+h)$ and $f(x)$ with their respective algebraic forms.
3. Simplify the substitutions algebraically.

Before we begin to cover exercises that involve the difference quotient, we note two things. First, the utility of the answers to these exercises may not be clear until later chapters in the book. Second, some business calculus texts use the notation $f(x+\Delta x)$ instead of $f(x+h)$. Use of both of these notations will be covered in the exercises.

EXAMPLE 1.8

Use the difference quotient to evaluate the following functions.

a) $f(x) = 3$

b) $f(x) = x+5$

c) $f(x) = x^2 - 2x+5$

d) $f(x) = x^3 - x$

e) $f(x) = \sqrt{x}$

f) $f(x) = \frac{1}{x}$

SOLUTION 1.8

a) $f(x) = 3$ — Copy the given function.

$f(x+h) = 3$ — Evaluate the given function at $x+h$.

$$\frac{f(x+h)-f(x)}{h} = \frac{3-3}{h}$$

Place the evaluation for $f(x)$ and $f(x+h)$ into the difference quotient formula.

$$\frac{3-3}{h} = \frac{0}{h} = 0$$

Use algebra to evaluate the formula.

The solution is 0.

b) $f(x) = x+5$ — Copy the given function.

$f(x+h) = (x+h)+5$ — Evaluate the given function by substituting $x+h$ for x.

$$\frac{f(x+h)-f(x)}{h} = \frac{[(x+h)+5]-(x+5)}{h}$$

Replace $f(x)$ with $x + 5$ and $f(x+h)$ with $(x+h)+5$ in the difference quotient.

$$\frac{x+h+5-x-5}{h} = \frac{h}{h} = 1$$

Distribute the minus sign through $(x + 5)$ and subtract in the numerator.

The evaluation of the difference quotient for this function is 1.

c) $f(x) = x^2-2x+5$ — Copy the given function.

$f(x+h) = (x+h)^2-2(x+h)+5$ — Evaluate the given function by substituting $(x+h)$ for x.

Note: $x + h$ is substituted everywhere there was the letter x. Just as x was squared, so must the binomial $(x + h)$ be squared.

$$\begin{aligned}(x+h)^2 &= (x+h)\cdot(x+h)\\ &= x^2+hx+hx+h^2\\ &= x^2+2hx+h^2\end{aligned}$$

Expand $(x+h)^2$

$$\frac{[(x^2+2hx+h^2)-2(x+h)+5]-[x^2-2x+5]}{h}$$

Replace $f(x+h)$ and $f(x)$ with their equivalent algebraic forms.

$$\frac{x^2 + 2hx + h^2 - 2x - 2h + 5 - x^2 + 2x - 5}{h}$$

Distribute –2 and the minus sign. Drop all grouping symbols.

$$\frac{2hx + h^2 - 2h}{h}$$

Combine like terms.

$$\frac{h(2x + h - 2)}{h}$$

Factor h out of the numerator.

$2x + h - 2$

Cancel the h factor in the numerator and the denominator.

$2x + h - 2$ is the result of the difference quotient for this given function.

In the next example, we will use $(x + \Delta x)$ instead of $(x + h)$.

d) $f(x) = x^3 - x$

Copy the given function.

$f(x + \Delta x) = (x + \Delta x)^3 - (x + \Delta x)$

Replace every x with the binomial $(x + \Delta x)$.

$(x + \Delta x)^3 = (x + \Delta x)(x + \Delta x)(x + \Delta x)$

$(x + \Delta x)^3$ is expanded. Δx is treated as a single element.

$= (x^2 + x\Delta x + x\Delta x + (\Delta x)^2)(x + \Delta x)$

$= (x^2 + 2x\Delta x + (\Delta x)^2)(x + \Delta x)$

$= x^3 + x^2(\Delta x) + 2x^2(\Delta x) + 2x(\Delta x)^2 + x(\Delta x)^2 + (\Delta x)^3$

$= x^3 + 3x^2(\Delta x) + 3x(\Delta x)^2 + (\Delta x)^3$

$$\frac{[x^3 + 3x^2(\Delta x) + 3x(\Delta x)^2 + (\Delta x)^3 - (x + \Delta x)] - (x^3 - x)}{\Delta x}$$

Replace $f(x + \Delta x)$ and $f(x)$ with their equivalent algebraic forms.

Note that Δx also replaces h in the denominator.

$$\frac{x^3 + 3x^2\Delta x + 3x(\Delta x)^2 + (\Delta x)^3 - x - \Delta x - x^3 + x}{\Delta x}$$

Expand the numerator.

$$\frac{3x^2\Delta x + 3x(\Delta x)^2 + (\Delta x)^3 - \Delta x}{\Delta x}$$ Combine like terms.

$$\frac{\Delta x(3x^2 + 3x\Delta x + (\Delta x)^2 - 1)}{\Delta x}$$ Factor Δx from every numerator term.

$$\frac{\cancel{\Delta x}(3x^2 + 3x\Delta x + (\Delta x)^2 - 1)}{\cancel{\Delta x}}$$ Cancel Δx from the numerator and denominator.

$3x^2 + 3x\Delta x + (\Delta x)^2 - 1$ is the solution to the difference quotient formula.

Note: Some students find it more difficult to work with $(x + \Delta x)$ than $(x + h)$. This may be due to the similarity of the symbols x and Δx and the uneasiness of treating Δx as a single symbol. If there is some confusion for you, replace Δx with h at the beginning of the problem. Complete the exercise, then replace each h with Δx at the end of the problem.

e) $f(x) = \sqrt{x}$ Copy the given function.

$f(x+h) = \sqrt{x+h}$ Replace x with $x+h$.

$$\frac{\sqrt{x+h} - \sqrt{x}}{h}$$ Complete the replacement of $f(x+h)$ and $f(x)$ into the difference quotient.

$$= \frac{\sqrt{x+h} - \sqrt{x}}{h} \cdot \frac{\sqrt{x+h} + \sqrt{x}}{\sqrt{x+h} + \sqrt{x}}$$ Rationalize the numerator.

$$= \frac{x + h - \sqrt{x}\sqrt{x+h} + \sqrt{x}\sqrt{x+h} - x}{h(\sqrt{x+h} + \sqrt{x})}$$ Multiply and combine like terms.

$$= \frac{x + h - x}{h(\sqrt{x+h} + \sqrt{x})} = \frac{h}{h(\sqrt{x+h} + \sqrt{x})}$$ Simplify and, again, combine like terms.

$$= \frac{\cancel{h}}{\cancel{h}(\sqrt{x+h} + \sqrt{x})}$$ Cancel the h factor from the numerator and denominator.

$\dfrac{1}{\sqrt{x+h}-\sqrt{x}}$ is the solution to the difference quotient. Its use will be clearer in Chapter 3.

f) $f(x) = \dfrac{1}{x}$ — Copy the given function.

$f(x+h) = \dfrac{1}{x+h}$ — Replace x with the binomial $x+h$.

$$\frac{\dfrac{1}{x+h}-\dfrac{1}{x}}{h}$$

Substitute for $f(x)$ and $f(x+h)$ in the difference quotient

$$\frac{\dfrac{x}{x}\cdot\dfrac{1}{x+h}-\dfrac{1}{x}\cdot\dfrac{x+h}{x+h}}{h}$$

Use the common denominator $x(x+h)$ to combine the numerator terms.

$$\frac{\dfrac{x-x-h}{x(x+h)}}{h}$$

Distribute the minus.

$$\frac{\dfrac{-h}{x(x+h)}}{h}$$

Combine like terms in the numerator.

$$\frac{\dfrac{-h}{x(x+h)}}{\dfrac{h}{1}} = \frac{-h}{x(x+h)}\cdot\frac{1}{h}$$

Convert h to $\dfrac{h}{1}$ and invert the denominator to multiply.

$$\frac{-\cancel{h}}{x(x+h)}\cdot\frac{1}{\cancel{h}} = \frac{-1}{x(x+h)}$$

The h factors cancel to form the solution.

$\dfrac{-1}{x(x+h)} = \dfrac{-1}{x^2+hx}$ is the solution of the difference quotient.

One final note is worth discussing. You may have noticed that we went to fairly extraordinary effort to ensure that a single h factor was not left in the solution denominator. The solution to $f(x) = \sqrt{x}$ and $f(x) = 1/x$ bear this out.

In order to evaluate the difference quotient, we must make sure that the solutions do not have division by zero (that is, if h were replaced by 0).

1.4 LINEAR FUNCTIONS

All of the functions of the form $f(x) = mx + b$ are called linear functions. These functions will show a straight line when graphed. In this format, m is the slope, and b is the y-intercept. There are various aspects of these straight line functions that are studied in depth. These aspects will be discussed in this section.

Slope

The "steepness" of a line is called the *slope* of the line. The slope of the line is the ratio of a segment of vertical movement of the line (*rise*) to the horizontal movement of the line (*run*).

$$\textbf{slope} = \frac{\text{rise}}{\text{run}}$$

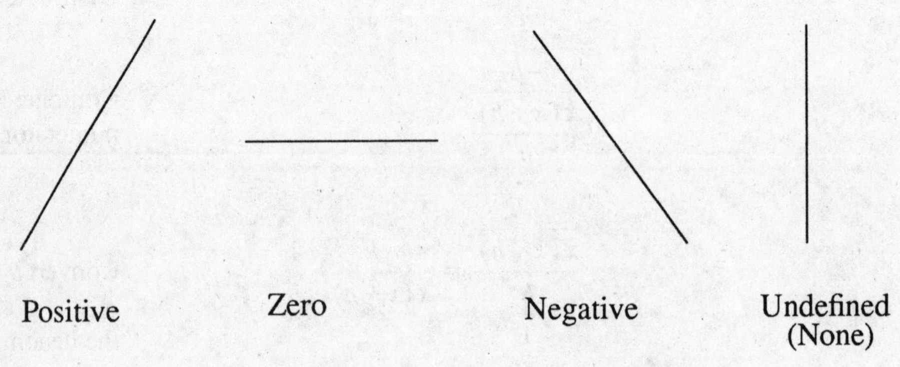

Figure 1.1 The Various Types of Slope

Figure 1.1 shows how the slope of the line relates to its slant. The less slant the line has, the smaller the absolute value of the slope. This figure shows that a line that is **horizontal** has a **slope of 0**. As the slant of the line climbs steeper as we read from left to right, the slope becomes larger positively.

Although the nuances of our language often allow the substitution of "zero" for "no," this cannot be the case when we talk about slopes. It is important to understand that a vertical line doesn't have a slope and a horizontal line has a slope equal to the number 0.

Since the rise on a graph is the difference between y-values, and the run on the graph is the difference between x-values, a slope can be determined from two points (x_1, y_1) and (x_2, y_2) on a graph by the following formula. The letter $\boldsymbol{m}$ is the mathematical designation for **slope**.

$$m = \frac{y_2 - y_1}{x_2 - x_1}$$

The line that connects point A, $(3, 2)$, and point B, $(-1, 4)$, has a slope defined by:

$$m = \frac{4-2}{-1-3} = \frac{2}{-4} = -\frac{1}{2}$$

Note: It does not matter which point is labeled point A or which is labeled point B; the slope that is computed will be the same.

EXAMPLE 1.9

Find the slope of the line passing through the following points. It does not matter which point is labeled (x_1, y_1) and which is labeled (x_2, y_2):

a) $(2, 1)$ and $(7, 1)$
b) $(2, 0)$ and $(-3, -1)$
c) $(1, -2)$ and $(1, 3)$

SOLUTION 1.9

a) $m = \frac{y_2 - y_1}{x_2 - x_1} = \frac{1-1}{7-2} = \frac{0}{5} = 0$

b) $m = \frac{y_2 - y_1}{x_2 - x_1} = \frac{0-(-1)}{2-(-3)} = \frac{1}{5}$

c) $m = \frac{y_2 - y_1}{x_2 - x_1} = \frac{3-(-2)}{1-1} = \frac{5}{0} = \text{undefined}$

Division by 0 is undefined.

Two formulas are often used to convert information about a line into an equation for that line. These two formulas are:

slope-intercept form	$y = mx + b$
where: m is the slope	
b is the y-intercept (where the line crosses the y-axis)	

point-slope form	$y - y_1 = m(x - x_1)$
where: m is the slope	
(x_1, y_1) are the coordinates of one point on the line	
x and y are not replaced.	

The formulas can be used to convert the information about the line into one of the equation forms. Some additional information will be helpful as we seek to formulate linear equations:

a) The slopes of two **parallel lines** with slopes of m_1 and m_2, respectively, are equal. That is, $m_1 = m_2$.
b) The slopes of two perpendicular lines are the negative reciprocals of each other. That is, $m_1 = \frac{-1}{m_2}$.
c) The coordinates of the origin are $(0, 0)$.
d) The point at which the line crosses the x-axis is the x-intercept. Its coordinates are (x-value, 0).
e) The points at which the line crosses the y-axis is the y-intercept. The coordinates for the y-intercept are (0, y-value).

Armed with the above information and the two formulas for developing the line equations, we can now find the equation of a line given various information.

EXAMPLE 1.10

a) Determine the equation of the line passing through the points $(2, 1)$ and $(-1, 3)$.
b) Find the equation of the line with slope of $\frac{3}{2}$ and a y-intercept of -2.
c) Write the equation of a line with x-intercept of 2 and passing through $(1, 5)$.
d) Find the equation of a line passing through the origin and crossing $(-7, 1)$.
e) Find the equation of a line with x-intercept of -5 and y-intercept of -2.
f) Formulate the equation of the line parallel to the line $2x - 4y = 3$ and passing through $(8, 0)$.

g) Determine the equation of a line perpendicular to $y = -\frac{1}{5}x + 1$ and passing through $(0, -1)$.

h) Find the equation of the horizontal line passing through $(-2, 3)$.

i) Write the equation of the vertical line crossing the x-axis at -7.

SOLUTION 1.10

a) With two points given, we can first find the slope and then use one of the points and the slope in the point-slope formula.

$(2, 1) \quad (-1, 3)$	Two points are given.
$m = \frac{y_2 - y_1}{x_2 - x_1} = \frac{3-1}{-1-2} = \frac{2}{-3} = -\frac{2}{3}$	Find the slope.
$m = \frac{-2}{3} \qquad (2, 1) = (x_1, y_1)$	Select either one of the points and use the slope found earlier.
$y - y_1 = m(x - x_1)$	We now need the point-slope formula.
$y - 1 = -\frac{2}{3}(x - 2)$	Substitute for slope and point.
$(3)(y-1) = (3)\left(-\frac{2}{3}\right)(x-2)$	Multiply both sides of the equation by the denominator of the slope.
$3(y-1) = -2(x-2)$	
$3y - 3 = -2x + 4$	Distribute to remove parentheses.
$3y - 3 = -2x + 4$ $\underline{\;\;+3 \qquad\quad +3}$ $3y \qquad = -2x + 7$	Add 3 to both sides.
$\underline{\;\;+2x \quad +2x}$	Add $2x$ to both sides.
$2x + 3y = 7$	This completes the equation and isolates the constant (7).

b) Knowing the slope $(m = 3/2)$ and the y-intercept $(b = -2)$, this problem is tailor-made for the slope-intercept formula.

$y = mx + b$	Write the slope-intercept form of the equation.
$y = \frac{3}{2}x + (-2)$	Substitute slope and intercept.
$y = \frac{3}{2}x - 2$	You could leave the equation in this form.

$$(2)y = 2\left(\frac{3}{2}x - 2\right)$$

Or you could change it into the standard form by multiplying by the denominator of the slope.

$$2y = 2\left(\frac{3}{2}x\right) - 2(2)$$

Multiply through by 2. Don't forget to distribute the 2 on the right through both terms.

$$2y = 3x - 4$$

$$-3x + 2y = -4$$

Isolate the constant term.

or

$$3x - 2y = 4$$

c) In this case the equation will be derived from two points (the x-intercept and the point $(1, 5)$).

The x-intercept of 2 is the point $(2, 0)$. Thus, the points this line passes through are $(2, 0)$ and $(1, 5)$.

$$m = \frac{y_2 - y_1}{x_2 - x_1} = \frac{5 - 0}{1 - 2} = \frac{5}{-1} = -5$$

Find the slope.

$$y - y_1 = m(x - x_1)$$

We now need the point-slope formula.

$$m = -5 \qquad (x_1, y_1) = (2, 0)$$

$$y - 0 = -5(x - 2)$$

Substitute into the formula.

$$y = -5(x - 2)$$

$$y = -5x + 10$$

Distribute.

$$5x + y = 10$$

Isolate the constant term.

d) This equation will also be built from two points. We will use the origin $(0, 0)$ and the point $(-7, 1)$.

$$m = \frac{y_2 - y_1}{x_2 - x_1} = \frac{1 - 0}{-7 - 0} = \frac{-1}{7}$$

Find the slope.

$$y - y_1 = m(x - x_1)$$

We now need the point-slope formula.

$$y - 0 = \frac{-1}{7}(x - 0)$$

Substitute into point-slope formula.

$$y = \frac{-1}{7}x$$

Simplify the equation.

$7y = 7\left(\frac{-1}{7}x\right)$	Multiply both sides of the equation by the denominator.
$7y = -x$	
$7y + x = -x + x$	Add x to both sides of the equation.
$x + 7y = 0$	

Note: The standard form of any line passing through the origin will be an equation of the form $ax + by = 0$.

e) This problem has two intercepts, which yield two points. The points are $(-5, 0)$ from the x-intercept of -5 and $(0, -2)$ from the y-intercept of -2.

$m = \frac{y_2 - y_1}{x_2 - x_1} = \frac{-2-0}{0-(-5)} = \frac{-2}{5}$	Find the slope.
$m = \frac{-2}{5} \qquad b = -2$	We could use the point-slope form, but the slope-intercept is easier.
$y = mx + b$	Write the slope-intercept form of the equation.
$y = \frac{-2}{5}x - 2$	Substitute into the equation.
$(5)y = (5)\left(\frac{-2}{5}x - 2\right)$	Multiply both sides of the equation by the denominator.
$5y = 5\left(\frac{-2}{5}x\right) - 5(2)$	Distribute.
$5y = -2x - 10$	
$5y + 2x = -2x + 2x - 10$	Add $2x$ to both sides.
$2x + 5y = -10$	Isolate the constant term.

f) We have a slope "hidden" in the line that is given as parallel to the line we are to find and we are given a point (8, 0).

$2x - 4y = 3$	Solve for y in terms of x to find the slope of the given line.
$-4y = -2x + 3$	

$$\frac{-4y}{-4} = \frac{-2x}{-4} + \frac{3}{-4}$$ Divide both sides by −4.

$$y = \frac{1}{2}x - \frac{3}{4}$$ The equation is now in slope-intercept form. We will use only the slope and disregard the rest of the linear equation.

$$(x_1, y_1) = (8, 0) \qquad m = \frac{1}{2}$$

$$y - y_1 = m(x - x_1)$$ Use the point-slope form.

$$y - 0 = \frac{1}{2}(x - 8)$$ Substitute for the slope and the point.

$$2(y - 0) = 2\left(\frac{1}{2}\right)(x - 8)$$ Multiply both sides by the denominator of the slope.

$$-x + 2y = -8$$ Subtract x from both sides.

or

$$x - 2y = 8$$

g) The slope is obtained from $y = -\frac{1}{5}x + 1$ and the point is $(0, -1)$. The line contains the slope needed for this problem. The slope of $y = \frac{-1}{5}x + 1$ is the coefficient of $x\left(-\frac{1}{5}\right)$. Perpendicular lines have slopes that are negative reciprocals of each other. We find the new slope by taking $-\frac{1}{5}$ and "flipping" it to get $\frac{5}{-1} = -5$. Next, we change the sign to get 5. The slope of the line perpendicular to the given line is 5.

$$(x_1, y_1) = (0, -1) \qquad m = 5$$

$$y - y_1 = m(x - x_1)$$ Use the point-slope form.

$$y - (-1) = 5(x - 0)$$ Substitute for slope and point.

$$y + 1 = 5x$$ Simplify.

$$y + 1 - 1 = 5x - 1$$ Subtract 1 from both sides.

$$y = 5x - 1$$

h) The equation of a horizontal line is $y = y$-value. Since we were given $(-2, 3)$, the equation must be $y = 3$.

i) The equation of a vertical line is $x = x$-value. Since we were given $(-7, 0)$, the equation must be $x = -7$.

1.5 SOLVING INEQUALITIES

Rules for Inequalities

While linear equations with one unknown have only one solution, linear inequalities usually have many (or a range) of solutions.

Symbol	Meaning
$x > y$	x is greater than y
$x < y$	x is less than y
$x \geq y$	x is greater than or equal to y
$x \leq y$	x is less than or equal to y

Table 1.1 Inequality Symbols

Rule	Format
addition	$a + c > b + c$
subtraction	$a - c > b - c$
multiplication	$a \cdot c > b \cdot c; \quad c > 0$
division	$a \div c > b \div c; \quad c > 0$
negation	$-a < -b$

Table 1.2 Inequality Rules for $a > b$

In Table 1.2, you might notice that a negation of the inequality reverses the direction (sometimes referred to as "the sense") of the inequality. That is, the > will reverse to < if the inequality is divided or multiplied by a negative number. Addition and subtraction do not affect the direction of the inequality.

EXAMPLE 1.11

Solve the following inequalities:

a) $x + 5 > 7$

b) $2x - 3 < 5$

c) $4 - 5x > 24$

d) $6 - \frac{x}{3} < 18$

SOLUTION 1.11

a) $x + 5 > 7$ — Copy the given inequality.

$$\begin{array}{r} x+5 > 7 \\ \underline{-5 \quad -5} \end{array}$$

Subtract 5 from each side of the inequality.

$$x > 2$$

Solution set is all real numbers greater than 2.

b) $2x - 3 < 5$ — Copy the given inequality.

$$\begin{array}{r} 2x-3 < 5 \\ \underline{+3 \quad +3} \\ 2x < 8 \end{array}$$

Add 3 to both sides of the inequality.

$$\frac{2x}{2} < \frac{8}{2}$$

Divide by 2.

$$x < 4$$

The solution set includes all real numbers has less than 4. Check by substituting numbers less than 4 into the original inequality.

c) $4 - 5x > 24$ — Copy the given inequality.

$$\begin{array}{r} 4-5x > 24 \\ \underline{-4 \qquad -4} \\ -5x > 20 \end{array}$$

Subtract 4 from each side of the inequality.

$$\frac{-5x}{-5} < \frac{20}{-5}$$

We divide by –5. Remember that when we divide by a negative number, we reverse the direction of the inequality.

$$x < -4$$

The direction of the inequality has been reversed. The solution set includes all numbers less than –4.

Check by substituting numbers less than –4 into the original inequality.

d) $6 - \frac{x}{3} < 18$ — Copy the given inequality.

$6 - \frac{x}{3} < 18$ — Subtract 6 from each side of the inequality.

$-\frac{x}{3} < 12$

$-\frac{3}{1} \cdot \frac{-x}{3} > (-3)\ 12$ — Multiply both sides of the inequality by (–3).

$\frac{-3x}{-3} > -36$ — The direction of the inequality is reversed.

$x > -36$ — The solution set is any number greater than –36. Check by placing any number larger than –36 in the original inequality.

Double Inequalities

Sometimes an inequality contains two inequality symbols. These types of problems involve a double inequality and may take the form:

$$a < x < b$$

The rules in Table 1.2 apply.

EXAMPLE 1.12

Find the solution to the following inequalities:

a) $-3 < x + 4 < 8$

b) $2 < 2x - 8 < 3$

c) $5 < 3 - x < 8$

d) $2 \le 4 - \frac{x}{2} \le 4$

SOLUTION 1.12

a) $-3 < x + 4 < 8$ — Copy the given inequality.

$$\begin{array}{r} -3 < x + 4 < 8 \\ \underline{-4 \quad\quad -4 \;\; -4} \end{array}$$

Subtract 4 from each part of the problem separated by the inequality symbols.

$-7 < x < 4$	The solution set is any number between –7 and 4. Check the answer by placing any number between –7 and 4 into the original double inequality.
b) $2 < 2x - 8 < 3$	Copy the given inequality.
$\begin{array}{l} 2 < 2x - 8 < 3 \\ \underline{+8 \quad\quad +8 \;\; +8} \\ 10 < 2x < 11 \end{array}$	Add 8 to each part of the inequality.
$\frac{10}{2} < \frac{2x}{2} < \frac{11}{2}$	Divide each part of the inequality by 2.
$5 < x < \frac{11}{2}$	Simplify.
$5 < x < \frac{11}{2}$	The solution set is any real number greater than 5 and less than 11/2. Check the answer by placing any number greater than 5 up to (but not including) 11/2 into the original inequality.
c) $5 < 3 - x < 8$	Copy the given inequality.
$\begin{array}{l} 5 < \;3 - x < \;8 \\ \underline{-3 \;\; -3 \quad\quad -3} \\ 2 < -x < 5 \end{array}$	Subtract 3 from all parts of the inequality.
$(-1)\,2 > (-1)\,(-x) > (-1)\,(5)$	Multiplying by the same negative number (–1) reverses the direction of the inequality.
$-2 > x > -5$	The solution set to the double inequality is any real number between –2 and –5. Check by using any number from the solution set in the original double inequality.

d) $2 \le 4 - \frac{x}{2} \le 4$ — Copy the given inequality.

$$\begin{array}{l} 2 \le 4 - \frac{x}{2} \le 4 \\ \underline{-4 \quad -4 \qquad -4} \end{array}$$

Subtract 4 from each side of the inequality.

$$-2 \le -\frac{x}{2} \le 0$$

$$2(-2) \le 2\left(-\frac{x}{2}\right) \le 2(0)$$

Multiply each part of the inequality by 2.

$$-4 \le -x \le 0$$

$$(-1)(-4) \ge (-1)(-x) \ge (-1)(0)$$

Multiply by (–1).

$$4 \ge x \ge 0$$

Reverse the direction of the inequality signs.

$$4 \ge x \ge 0$$

The solution set is any real number between 0 and 4. Check the solution by placing any number from the solution set in the original double inequality.

1.6 ABSOLUTE VALUES IN EQUATIONS

Introduction

If a variable is placed within absolute value symbols and set equal to a number, two values for the variable will make the equation true. For example,

$$|x| = 3$$

gives two values for x, 3 and –3. This is true because $|-3| = 3$ and $|3| = 3$.

One method of solving absolute value equations is to form two separate equations from the absolute value. One equation will use the positive form of the expression inside the absolute value symbols, the other equation will use the negative form of the expression inside the absolute value symbols. For example, the expression inside the absolute value symbols in $|x+7| = 5$ is $x+7$. The positive form of this expression is $+(x+7)$. The negative form is $-(x+7)$. Both of these forms are set equal to 5. Therefore, we have:

$$\begin{array}{lcr} x+7 = 5 & \text{and} & -(x+7) = 5 \\ x = -2 & & -x-7 = 5 \\ & & -x = 12 \\ & & x = -12 \end{array}$$

We used the rules of solving equations to find the 2 solutions, –2 and –12.

EXAMPLE 1.13

Solve the following absolute value equations:

a) $|x+2| = 3$

b) $|2x-7| = -8$

c) $\left|\frac{x}{3}+2\right| = 1$

d) $|5-x| = 4$

SOLUTION 1.13

a) $|x+2| = 3$	Copy the given equation.
$(i)\ (x+2) = 3 \quad (ii)\ -(x+2) = 3$	The positive form of expression (*i*) and the negative form of expression (*ii*) are set equal to 3.
$(i)\ x+2 = 3 \quad (ii)\ -x-2 = 3$	Distribute -1 through all terms of the negative form.
$(i)\ x+2=3,\ -2\ -2,\ x = 1 \quad (ii)\ -x-2=3,\ +2\ +2,\ -x = 5$	Subtract 2 from (*i*) and add 2 to (*ii*).
$(-1)(-x) = (-1)(5)$; $x = -5$	Multiply by -1.
$|x+2| = 3,\ |1+2| = 3,\ |3| = 3,\ 3 = 3 \quad |x+2| = 3,\ |-5+2| = 3,\ |-3| = 3,\ 3 = 3$	Place each solution in the original absolute value to check the solution.
b) $|2x-7| = -8$	Copy the given equation.
$(i)\ 2x-7 = -8 \quad (ii)\ -(2x-7) = -8$	The positive form of expression (*i*) and the negative form of expression (*ii*) are set equal to -8.
$(i)\ 2x-7 = -8 \quad (ii)\ -2x+7 = -8$	Distribute -1 through all terms of the negative form.
$(i)\ 2x-7=-8,\ +7\ +7,\ 2x = -1 \quad (ii)\ -2x+7=-8,\ -7\ -7,\ -2x = -15$	Add 7 to (*i*) and subtract 7 from (*ii*).
$(i)\ \frac{2x}{2} = -\frac{1}{2} \quad (ii)\ \frac{-2x}{-2} = \frac{-15}{-2}$	Divide (*i*) by 2 and divide (*ii*) by -2.

(*i*) $x = -\frac{1}{2}$ (*ii*) $x = \frac{15}{2}$

$|2x - 7| = -8$ — Place each solution in the original absolute value to check the solution.

$\left|2\left(-\frac{1}{2}\right) - 7\right| = -8$

$|-1 - 7| \neq -8$ — $-1/2$ does not check.

$\left|2\left(\frac{15}{2}\right) - 7\right| = -8$

$|15 - 7| \neq -8$ — $15/2$ does not check.

Note that since neither solution checks, there is no solution to the given absolute value equation. In fact, no absolute value can equal a negative number and, thus, an equation of the form $|x| = -a$ for $a > 0$ has no solution.

c) $\left|\frac{x}{3} + 2\right| = 1$ — Copy the given equation.

(*i*) $\frac{x}{3} + 2 = 1$ (*ii*) $-\left(\frac{x}{3} + 2\right) = 1$ — The positive form of expression (*i*) and the negative form of expression (*ii*) are set equal to 1.

(*i*) $\frac{x}{3} + 2 = 1$ (*ii*) $-\frac{x}{3} - 2 = 1$ — Distribute -1 through all terms of the negative form.

(*i*) $\frac{x}{3} + 2 = 1$ (*ii*) $-\frac{x}{3} - 2 = 1$ — Subtract 2 from (*i*) and add 2 to (*ii*).

$\quad -2 \quad -2 \qquad\qquad +2 \quad +2$

(*i*) $\frac{x}{3} = -1$ (*ii*) $-\frac{x}{3} = 3$

(*i*) $3 \cdot \frac{x}{3} = -1\,(3)$

(*ii*) $-3 \cdot -\frac{x}{3} = 3\,(-3)$ — Multiply (*i*) by 3 and (*ii*) by -3.

$x = -3 \qquad x = -9$ — Check the solutions by placing them in the original absolute value equation.

d) $|5 - x| = 4$ — Copy the given equation.

(*i*) $5 - x = 4$ (*ii*) $-(5 - x) = 4$ — The positive form of expression (*i*) and the negagive form of expression (*ii*) are set equal to 4.

(*i*) $5 - x = 4$ (*ii*) $-5 + x = 4$

Distribute -1 through all terms of the negative form (*ii*).

(*i*) $5 - x = 4$ (*ii*) $-5 + x = 4$

$\underline{-5 \quad -5}$ $\underline{+5 \quad +5}$

$-x = -1$ $x = 9$

Subtract 5 from (*i*) and add 5 to (*ii*).

$(-1)(-x) = (-1)(-1)$

$x = 1$

Multiply (*i*) by -1. Check the solutions by placing them in the original absolute value equation.

1.7 ABSOLUTE VALUE INEQUALITIES

When absolute values are used in linear inequalities, there are generally two types of results. One result gives two sets of numbers, while the other gives one set with real number endpoints.

The solution process for absolute value inequalities uses the techniques for solving absolute value equations combined with inequality rules. This process will be clarified through the use of examples.

EXAMPLE 1.14

Solve the following absolute value inequalities:

a) $|x - 5| > 2$
b) $|4 - x| < 3$
c) $|2x - 1| < 6$
d) $\left|7 - \frac{x}{3}\right| > 1$

SOLUTION 1.14

a) $|x - 5| > 2$

Copy the given inequality.

(*i*) $(x - 5) > 2$ (*ii*) $-(x - 5) > 2$

The positive form of expression (*i*) and the negative form of expression (*ii*) are set greater than 2.

(*i*) $x - 5 > 2$ (*ii*) $-x + 5 > 2$

Distribute -1 through the left side of (*ii*).

(*i*) $x - 5 > 2$ (*ii*) $-x + 5 > 2$ $\quad +5 \; +5$ $\quad -5 \; -5$ $x \quad > 7$ $-x \quad > -3$	Add 5 to (*i*) and subtract 5 from (*ii*).
$(-1)\,(-x) < (-1)\,(-3)$ $x < 3$	Multiply by -1 and reverse the direction of the inequality symbol in (*ii*).
$x > 7$ or $x < 3$	Overall then, two sets of numbers are the solutions. Check by placing any members of these sets in the original absolute value inequality.
b) $\lvert 4 - x \rvert < 3$	Copy the given inequality.
(*i*) $(4 - x) < 3$ (*ii*) $-(4 - x) < 3$	The positive form of expression (*i*) and the negative form of expression (*ii*) are set less than 3.
(*i*) $4 - x < 3$ (*ii*) $-4 + x < 3$	Distribute -1 through the left side of (*ii*).
(*i*) $4 - x < 3$ (*ii*) $-4 + x < 3$ $-4 \quad\quad -4$ $+4 \quad\quad +4$ $-x < -1$ $x < 7$	Subtract 4 from (*i*) and add 4 to (*ii*).
(*i*) $(-1)\,(-x) > (-1)\,(-1)$	Multiply through by -1 in (*i*).
(*i*) $x > 1$	Reverse direction of symbol in inequality.
Thus, $1 < x < 7$ is the solution.	Check the solution set by selecting a member of the set and placing it in the original inequality.
c) $\lvert 2x - 1 \rvert < 6$	Copy the given inequality.
(*i*) $2x - 1 < 6$ (*ii*) $-(2x - 1) < 6$	The positive form of expression (*i*) and the negative form of expression (*ii*) are set less than 6.
(*i*) $2x - 1 < 6$ (*ii*) $-2x + 1 < 6$	Distribute -1 through the left side of (*ii*).

$$(i)\ \begin{array}{r} 2x - 1 < 6 \\ +1 \quad +1 \\ \hline 2x \quad < 7 \end{array} \qquad (ii)\ \begin{array}{r} -2x + 1 < 6 \\ -1 \quad -1 \\ \hline -2x \quad < 5 \end{array}$$

Add 1 to (i) and subtract 1 from (ii).

$$(i)\ \frac{2x}{2} < \frac{7}{2} \qquad (ii)\ -\frac{2x}{2} < \frac{5}{2}$$

$$x < \frac{7}{2} \qquad -x < \frac{5}{2}$$

Divide both inequalities by 2.

$$(-1)\ (-x) > (-1)\left(\frac{5}{2}\right)$$

$$x > -\frac{5}{2}$$

Multiply the inequality by –1 to reverse the direction of the inequality symbol.

The range, $-\frac{5}{2} < x < \frac{7}{2}$, is the solution.

Check this solution set by substituting a member of the set in the original inequality.

d) $\left|7 - \frac{x}{3}\right| > 1$

Copy the given inequality.

$$(i)\ 7 - \frac{x}{3} > 1 \qquad (ii)\ -\left(7 - \frac{x}{3}\right) > 1$$

The positive form of expression (i) and the negative form of expression (ii) are set greater than 1.

$$(i)\ 7 - \frac{x}{3} > 1 \qquad (ii)\ -7 + \frac{x}{3} > 1$$

Distribute –1 through left side of (ii).

$$(i)\ \begin{array}{r} 7 - \frac{x}{3} > 1 \\ \underline{-7 \qquad -7} \end{array} \qquad (ii)\ \begin{array}{r} -7 + \frac{x}{3} > 1 \\ \underline{+7 \qquad +7} \end{array}$$

$$-\frac{x}{3} > -6 \qquad \frac{x}{3} > 8$$

Subtract 7 from both sides of (i) and add 7 to both sides of (ii).

$$(i)\ 3\left(-\frac{x}{3}\right) > 3(-6) \qquad (ii)\ 3\left(\frac{x}{3}\right) > 3(8)$$

$$-x > -18 \qquad x > 24$$

Multiply both inequalities by 3 to clear fractions.

$$(i)\ (-1)\ (-x) < (-1)\ (-18)$$

Inequality (i) is multiplied by –1.

$$x < 18$$

Reverse the direction of the inequality symbol.

There are two solution sets:
(1) x-values less than 18.
(2) x-values greater than 24.

Check the solution by taking a member of each set and placing it in the original inequality.

1.8 GRAPHING EQUATIONS, INEQUALITIES, AND ABSOLUTE VALUE INEQUALITIES ON THE NUMBER LINE

The Number Line

The number line is a horizontal line that represents the set of all real numbers. Later in the text, it will be called the "x-axis."

A number line can be used to visually show the number, or set of numbers, that is the solution to an equation or an inequality. The number 0 is placed at a location on the number line called the "origin." To the left of the origin are the negative numbers. Numbers to the right are positive numbers. (See Figure 1.2.)

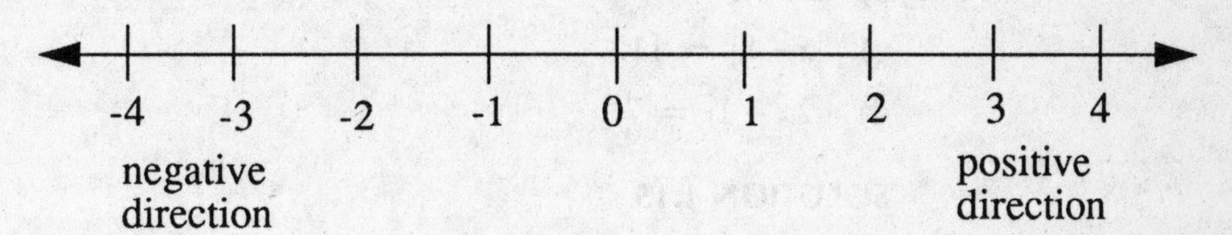

Figure 1.2 The Number Line

Each end goes on indefinitely.

Showing the Solution of an Equation on the Number Line

If the solution to an equation is $x = 2$, then a dot at 2 on the number line would represent the solution set for this equation (see Figure 1.3).

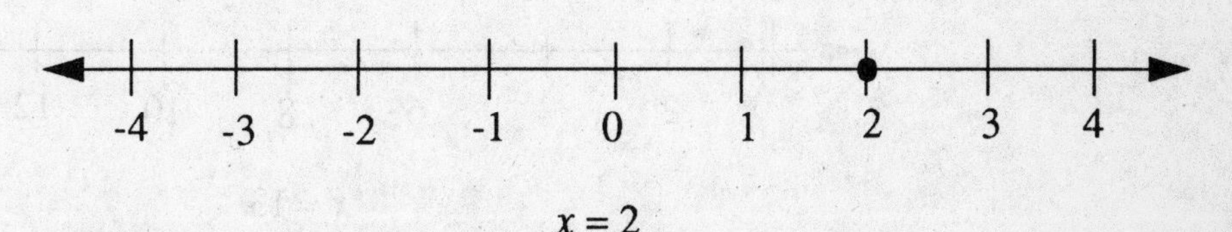

Figure 1.3 The Solution Set of $x = 2$

Showing the Solution of an Inequality on a Number Line

For an inequality, the solution set is a range of numbers. This can be indicated by a solid line segment on the number line that has a parenthesis or brace at the end of the line segment.

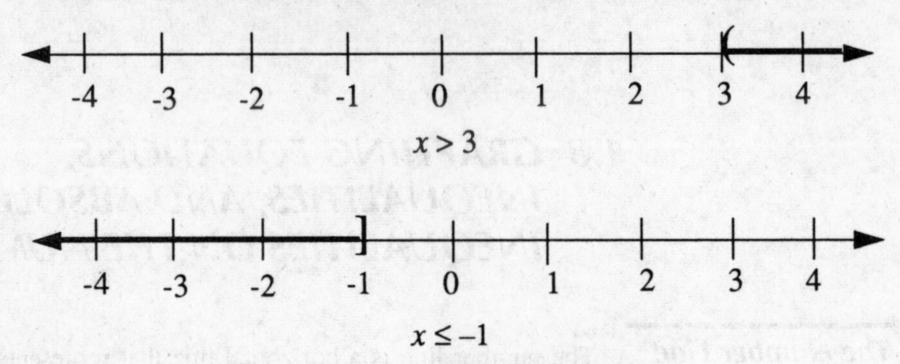

Figure 1.4 Examples of Inequalities

EXAMPLE 1.15

Solve the following equations and show their solution(s) on the number line.

a) $x - 5 = 8$

b) $2x + 5 = 4$

c) $|x - 8| = 11$

d) $|2x - 1| = 7$

SOLUTION 1.15

a) $x - 5 = 8$ — Copy the given equation.

$$\begin{array}{rcl} x - 5 & = & 8 \\ \underline{+5} & & \underline{+5} \\ x & = & 13 \end{array}$$

Add 5 to both sides of the equation.

The solution to the equation is 13. Check by placing the solution in the original equation.

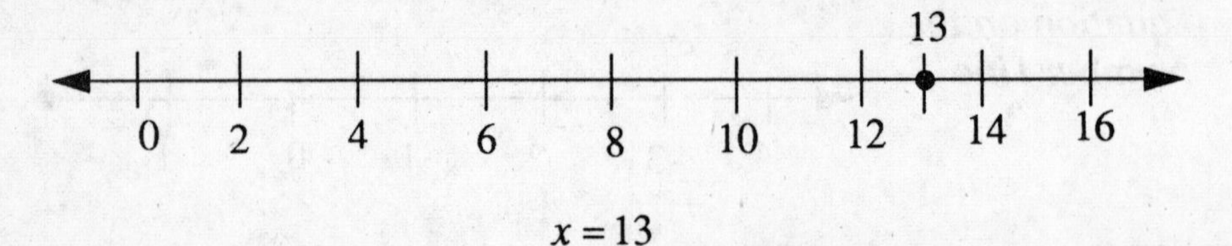

$x = 13$

Place a mark at 13 on the number line.

b) $2x + 5 = 4$ — Copy the given equation.

$$\begin{array}{rcl} 2x + 5 &=& 4 \\ -5 && -5 \\ \hline 2x &=& -1 \end{array}$$

Subtract 5 from both sides of the equation.

$$\frac{2x}{2} = \frac{-1}{2}$$

Divide by 2.

$$x = \frac{-1}{2}$$

Check the solution by placing it in the original equation.

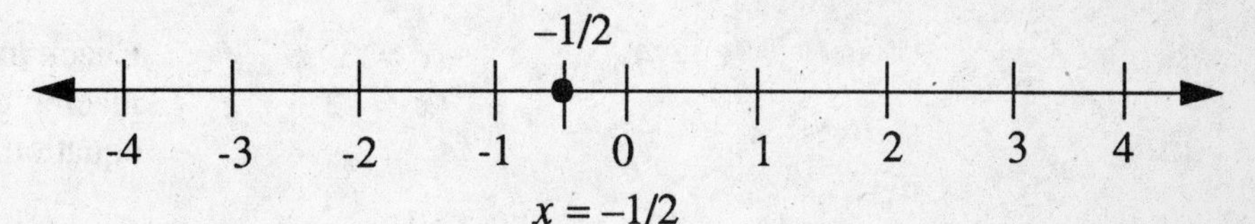

$x = -1/2$

Place a mark at the approximate location of $-\frac{1}{2}$ on the number line.

c) $|x - 8| = 11$ — Copy the given equation.

$(i)\ (x - 8) = 11 \quad (ii)\ -(x - 8) = 11$ — Separate the absolute value into two equations.

$(i)\ x - 8 = 11 \quad (ii)\ -x + 8 = 11$ — Distribute –1 through the left side of (*ii*).

$$\begin{array}{rcl} (i)\ x - 8 &=& 11 \\ +8 && +8 \\ \hline x &=& 19 \end{array} \qquad \begin{array}{rcl} (ii)\ -x + 8 &=& 11 \\ -8 && -8 \\ \hline -x &=& 3 \end{array}$$

Add 8 to both sides of (*i*) and subtract 8 from both sides of (*ii*).

$$(-1)\ (-x) = (-1)\ (3)$$

Multiply by –1 through equation (*ii*).

$$x = -3$$

Check by placing the solutions in the original absolute value equation.

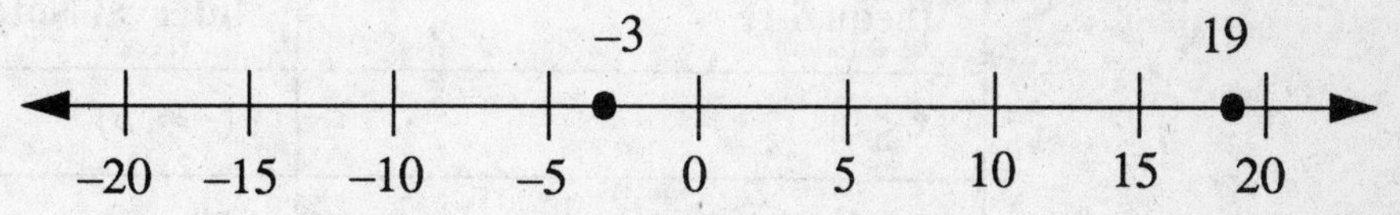

$x = -3,\ x = 19$

Mark both locations (–3) and (19) on the number line.

d) $\lvert 2x-1\rvert = 7$	Copy the given equation.
(*i*) $(2x-1)=7$ (*ii*) $-(2x-1)=7$	Separate the absolute value into two equations.
(*i*) $2x-1 = 7$ (*ii*) $-2x+1 = 7$	Distribute –1 through the left side of (*ii*).
(*i*) $2x-1 = 7$ (*ii*) $-2x+1 = 7$ $\quad +1 \quad +1 \qquad\qquad -1 \quad -1$ $2x = 8 \qquad -2x = 6$	Add 1 to (*i*) and subtract 1 from (*ii*).
$\frac{2x}{2} = \frac{8}{2} \qquad -\frac{2x}{2} = \frac{6}{2}$	Divide both equations by 2.
$x = 4 \qquad -x = 3$ $x = -3$	Check the solutions by placing them in the original equation.

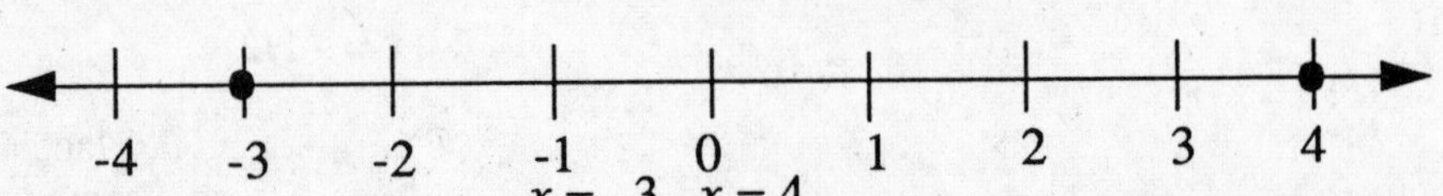

$x=-3,\ x=4$

Mark the solutions (–3) and (4) on the number line.

There is an additional item to consider when interval notation is used. The fact that the number line stretches from minus infinity $(-\infty)$ to plus infinity $(+\infty)$ is important in the use of interval notation symbols.

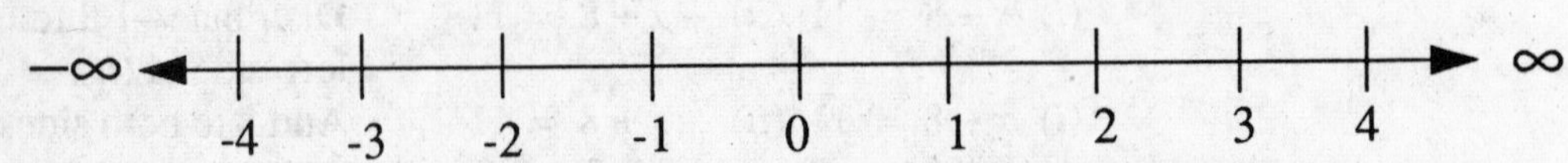

Figure 1.5 Showing Endpoints of Infinity

Parentheses are always used next to $(-\infty)$ or $(+\infty)$ because, theoretically, we never reach infinity. If we have an interval that includes the endpoint, we use a brace ([or]).

The following table lists some possibilities for interval notation.

Inequality	**Interval Notation**
$x < b$	$(-\infty, b)$
$x > b$	(b, ∞)
$x \leq b$	$(-\infty, b]$

Table 1.3 Interval Notation

Inequality	Interval Notation
$x \geq b$	$[b, \infty)$
$a \leq x \leq b$	$[a, b]$
$a < x < b$	(a, b)

Table 1.3 Interval Notation (con't.)

EXAMPLE 1.16

Use interval notation to describe the solutions of the following inequalities. Graph the inequality.

a) $2x + 3 > 11$
b) $5 - x \geq 4$
c) $|x + 2| < 6$
d) $|3x - 1| \geq 2$

SOLUTION 1.16

a) $2x + 3 > 11$ — Copy the given inequality.

$$\begin{array}{r} 2x + 3 > 11 \\ -3 \quad -3 \\ \hline 2x \quad > 8 \end{array}$$

Subtract 3 from both sides of the inequality.

$$\frac{2x}{2} > \frac{8}{2}$$

Divide by 2.

$x > 4$ — The solution set is all values greater than 4. Use "(" because the solution set does not contain 4.

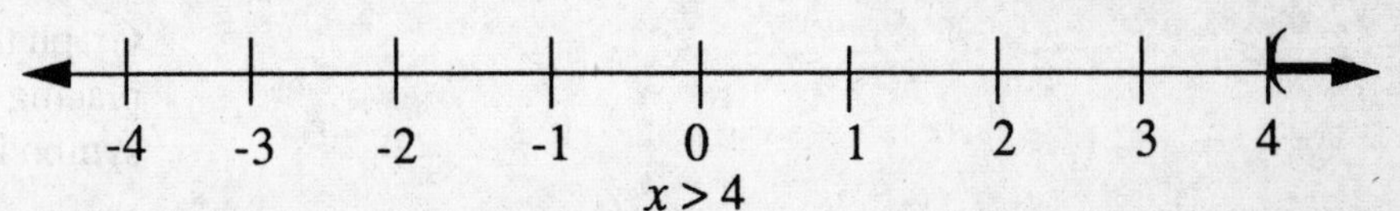

$x > 4$

$(4, \infty)$ — Express the solution set in interval notation.

b) $5 - x \geq 4$ — Copy the given inequality.

$$\begin{array}{r} 5 - x \geq 4 \\ -5 \quad\quad -5 \\ \hline -x \geq -1 \end{array}$$

Subtract 5 from both sides of the inequality.

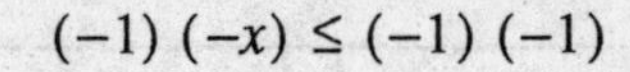

$(-1)\,(-x) \leq (-1)\,(-1)$ — Multiply by (–1) and reverse direction of the inequality.

$x \leq 1$ — Graph the inequality using "]" because the set contains 1 as well as those numbers less than 1.

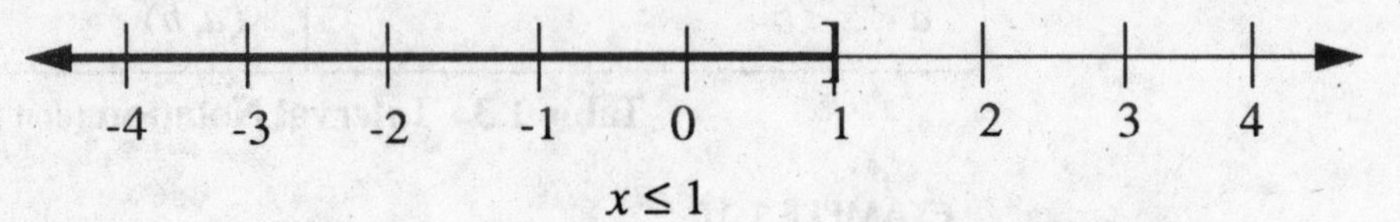

$x \leq 1$

$(-\infty, 1]$ — Write the solution set in interval notation.

c) $|x+2| < 6$ — Copy the given inequality.

(*i*) $(x+2) < 6$ (*ii*) $-(x+2) < 6$ — Separate the inequality into two inequalities.

(*i*) $x+2<6$ (*ii*) $-x-2<6$ — Distribute –1 through the left side of (*ii*).

(*i*) $x+2<6$ (*ii*) $-x-2<6$ — Subtract 2 from (*i*) and add 2 to (*ii*).

$$\begin{array}{lr} x+2 & <6 \\ -2 & -2 \\ \hline x & <4 \end{array} \qquad \begin{array}{lr} -x-2 & <6 \\ +2 & +2 \\ \hline -x & <8 \end{array}$$

$(-1)(-x) > (-1)(8)$ — Multiply (*ii*) by (–1).

$x > -8$ — Reverse the direction of the inequality.

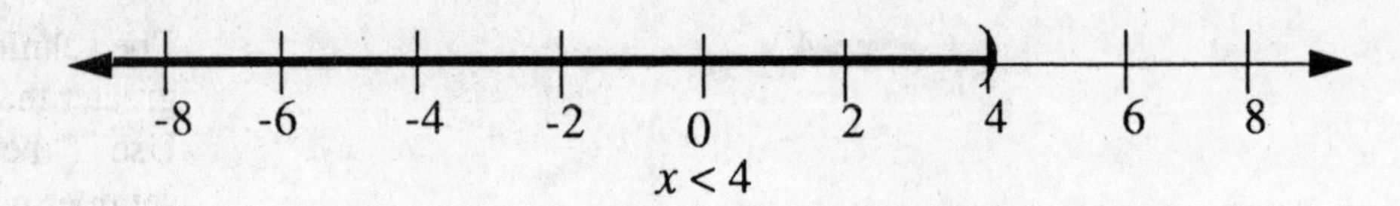

$x < 4$

Graph the solution set by placing a ")" at 4. This symbolizes $x < 4$.

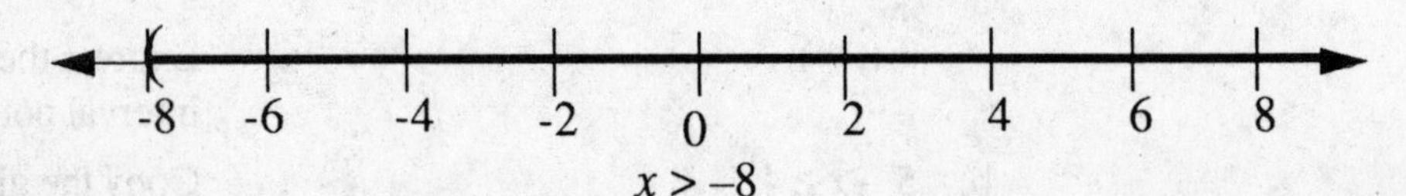

$x > -8$

Graph the condition that $x > -8$. Use "(" to indicate $x > -8$.

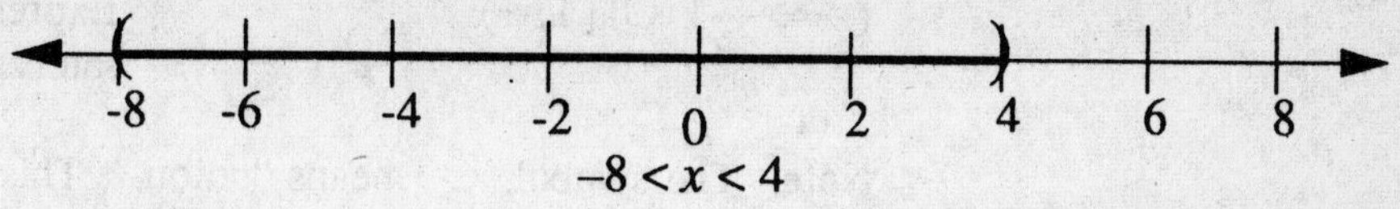

$-8 < x < 4$

$(-8, 4)$ Use interval notation to symbolize the solution set.

d) $|3x - 1| \geq 2$ — Copy the given equation.

(*i*) $(3x - 1) \geq 2$ (*ii*) $-(3x - 1) \geq 2$ — Separate the absolute value inequality into two separate inequalities.

(*i*) $3x - 1 \geq 2$ (*ii*) $-3x + 1 \geq 2$ — Distribute -1 across the left side of (*ii*).

(*i*) $3x - 1 \geq 2$ (*ii*) $-3x + 1 \geq 2$ — Add 1 to (*i*) and subtract 1 from (*ii*).

$$\begin{array}{rr} +1 & +1 \\ \hline 3x & \geq 3 \end{array} \qquad \begin{array}{rr} -1 & -1 \\ \hline -3x & \geq 1 \end{array}$$

(*i*) $\dfrac{3x}{3} \geq \dfrac{3}{3}$ (*ii*) $\dfrac{-3x}{3} \geq \dfrac{1}{3}$ — Divide by 3.

$x \geq 1$ $\qquad -x \geq \dfrac{1}{3}$

(*ii*) $(-1)(-x) \leq (-1)\left(\dfrac{1}{3}\right)$ — Multiply (*ii*) by (–1) and reverse the direction of the inequality.

$x \leq -\dfrac{1}{3}$

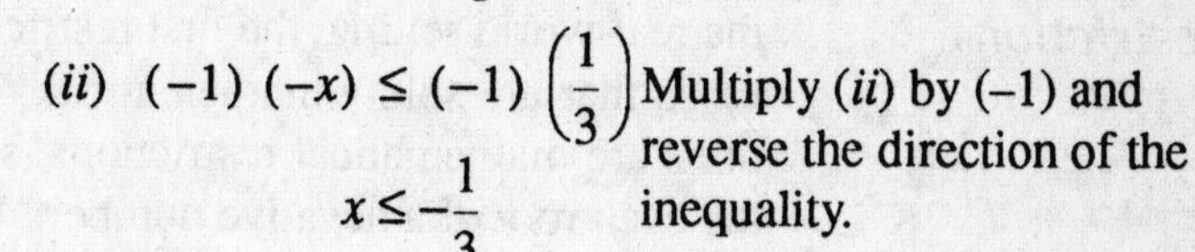

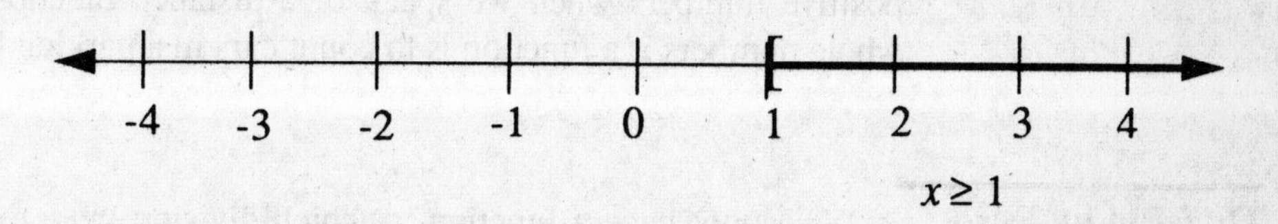

$x \geq 1$

Graph $x \geq 1$, using "[" because 1 is included in the solution set.

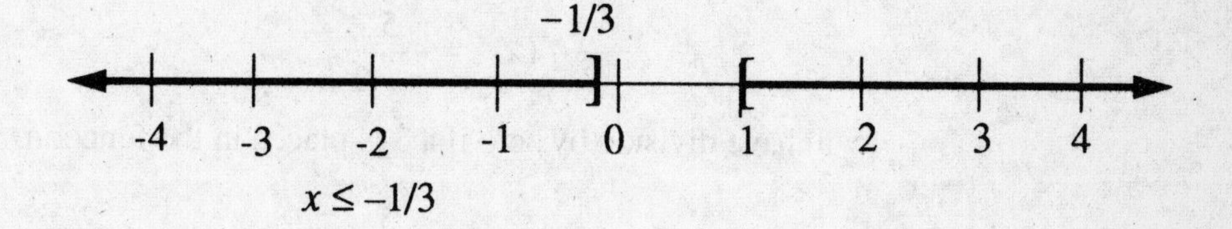

$x \leq -1/3$

Graph $x \leq -\dfrac{1}{3}$ using "]" because $-\dfrac{1}{3}$ is part of the solution set.

$(-\infty, -\frac{1}{3}] \cup [1, \infty)$ Express solution set in interval notation.

Note: The symbol, $\cup$, means "union." This symbol indicates that the separate intervals form one solution set.

1.9 RESTRICTIONS ON THE DOMAIN

The Meaning of Restrictions

As mentioned earlier, functions are operations of some form. We have not mentioned that, normally, the domain for a function is all real numbers. However, some real numbers will cause problems for a function so that it cannot be evaluated properly. Thus, the domain must be **restricted** so that those real numbers that cause problems for a given function will not be included in the set of domain elements.

Two Types of Restrictions

Since mathematics textbooks sometimes do not work with functions in the real-world setting, the first restrictions on a domain we will look at are those that are solely mathematical. The only restrictions on these functions are mathematical restrictions, such as: (1) division by zero and (2) the even root of a negative number. We will look at these two mathematical restrictions on the domain in this section. (Later on, we will talk about restrictions specific to the real-world use of a function, such as using only positive numbers when we speak of a distance function, or using only whole numbers if a function is to count cars in a parking lot.)

Division by Zero

We cannot have a function in which division by zero occurs. Therefore, if a number causes a zero to appear in the denominator, that number is restricted from the domain. For example, the function

$$f(x) = \frac{5}{x-3}$$

will have division by zero if a 3 is placed in the function.

$$f(x) = \frac{5}{x-3}$$

$$f(x) = \frac{5}{3-3} = \frac{5}{0}$$

For the above function, 3 is excluded from the domain. **All other** real numbers are valid domain elements.

It is often useful to determine ahead of time what real numbers are restricted from the domain.

> When division by zero is possible, set the denominator of the function equal to zero and solve. All solutions are restricted from the domain.

Even Roots of Negative Numbers

Another restriction on a domain occurs when a numerical substitution causes the function to try to find the even root of a negative number. Symbolically, this can be expressed as $f(a) = \sqrt[n]{a}$, where n is a positive even number and $a < 0$. When we try to find an even root, we try to find the square root, fourth root, sixth root, etc., of a number or expression. The difficulty that even roots have with negative numbers is best exemplified by looking at the square root. When we look for the square root of 9 ($\sqrt{9}$), we ask what number—multiplied times itself—would yield a product of 9. There are two answers to this: +3 and –3.

$$(+3)\,(+3) = +3 \cdot 3 = +9$$
$$(-3)\,(-3) = (-)\,(-)\,(3)\,(3) = +\,(9) = +9$$

The product of a negative times a negative is a positive.

If we search for the square root of –9 ($\sqrt{-9}$), we can find no real number that multiplies times itself to yield –9.

Some readers may know that a set of numbers called imaginary numbers have been defined to handle even roots of negative numbers. However, this set falls outside the set of real numbers and is not used in business calculus.

> To find the restriction on the domain caused by the even root of a negative number, set the contents of the radical expression < 0 and solve. This will produce the restricted numbers.

EXAMPLE 1.17

Find the restrictions on the domain for the following functions. Express the domain in interval notation.

a) $f(x) = \dfrac{10}{x-2}$

b) $f(x) = \sqrt{x+5}$

c) $f(x) = \dfrac{3x}{x^2+4x+3}$

d) $f(x) = \sqrt{3-2x}$

e) $f(x) = \dfrac{3-x}{5}$

f) $f(x) = 2x - \sqrt{3}$

g) $f(x) = \dfrac{\sqrt{x+2}}{4x-1}$

h) $f(x) = \dfrac{x-2}{\sqrt{2x+6}}$

i) $f(x) = \dfrac{4-x}{x^2+4}$

j) $f(x) = \sqrt[3]{x+1}$

SOLUTION 1.17

a) $f(x) = \dfrac{10}{x-2}$ — Copy the given function.

$x - 2 = 0$ — Restrictions on the domain occur when the denominator equals zero.

$x = 2$ — The real number, 2, must be excluded from the domain.

domain = $(-\infty, 2) \cup (2, \infty)$

b) $f(x) = \sqrt{x+5}$ — Copy the given function.

$x+5<0$ — The function has an even root. Set contents of radical < 0 to find exclusions on the domain.

$x<-5$

domain = $[-5, \infty)$ — All x-values **less than** –5 must be excluded. Therefore, –5 and all greater numbers are in the domain.

c) $f(x) = \dfrac{3x}{x^2+4x+3}$ — Copy the given function.

x^2+4x+3 — This function has an expression in the denominator, thus a possibility of division by 0.

$x^2+4x+3=0$ — Set the denominator equal to 0 and solve.

$(x+1)(x+3) = 0$ — Factor and solve by setting the factors equal to zero.

$x+1 = 0 \qquad x+3 = 0$

$x = -1 \qquad x = -3$

domain = $(-\infty, -3) \cup (-3, -1) \cup (-1, \infty)$

–1 and –3 are excluded from the domain. **All** other real numbers are members of the domain.

d) $f(x) = \sqrt{3-2x}$ — Copy the given function.

$3-2x<0$ — This function has an even root. Set contents of radical < 0 to find exclusions on the domain.

$-2x<-3$

$x > \dfrac{-3}{-2}$

$x > \dfrac{3}{2}$ — All x-values greater than $3/2$ must be excluded. This means that all real numbers **less than** or equal to $3/2$ are in the domain.

domain = $\left(-\infty, \dfrac{3}{2}\right]$

e) $f(x) = \dfrac{3-x}{5}$ — Copy the given function.

$5 \neq 0$ — Although this function has a denominator, that denominator cannot be set equal to 0. ($5 \neq 0$.)

no restrictions — The domain is all real numbers. This function has no mathematical restrictions on its domain.

f) $f(x) = 2x - \sqrt{3}$ — Copy the given function.

no restrictions — This function has no restrictions since the radical contains **only** the number 3, and 3 is not less than zero. The domain is all real numbers.

g) $f(x) = \dfrac{\sqrt{x+2}}{4x-1}$ — Copy the given function.

i. $4x - 1$ — This function has both a denominator and a radical expression. We first find exclusion due to division by zero. $1/4$ is excluded from the domain.

$$4x - 1 = 0$$
$$4x = 1$$
$$x = \frac{1}{4}$$

ii. $\sqrt{x+2}$ — Secondly, find the exclusion due to the radical expression.

$$x + 2 < 0$$
$$x < -2$$

All numbers less than -2 are excluded from the domain.

$$\text{domain} = \left[-2, \frac{1}{4}\right) \cup \left(\frac{1}{4}, \infty\right)$$

The domain contains all real numbers greater than or equal to -2, except for $1/4$.

h) $f(x) = \dfrac{x-2}{\sqrt{2x+6}}$ — Copy the given function.

$$2x + 6 < 0$$

$$2x < -6$$

In this instance, the radical expression is in the denominator. Normally it is possible for the expression

$x < -3$	in the radical to equal zero (because zero is not negative). However, zero in this case will cause division by 0. The domain = $(-3, \infty)$. The domain includes all numbers greater than –3. The number, –3, causes division by zero.
$(-3, \infty)$	Numbers below –3, inclusive, are excluded.
i) $f(x) = \dfrac{4-x}{x^2+4}$	Copy the given function.
$x^2 + 4 = 0$ $x^2 = -4$	If exclusions are to exist, the denominator must be set equal to 0 and the resulting equation solved.
$x = \sqrt{-4}$	The attempt to solve this equation yields an imaginary solution. Thus, this denominator can't equal zero.
$x^2 + 4 = 0$	A positive number plus four can't equal zero.

This function has no restrictions on the domain. Note, also, that the **numerator** equaling zero creates no problem for the function.

j) $f(x) = \sqrt[3]{x+1}$	Copy the given function.
$\sqrt[3]{x+1}$	This function contains an odd index (root). Odd roots have no restrictions with neither **positive** nor **negative** roots.
no restrictions	There are no restrictions on the domain.

1.10 LOGARITHMS

Logarithmic and Exponential Forms

There are occasions when we must solve for an unknown exponent or rewrite products as additions and quotients as differences. On these occasions we may need to use logarithms. Logarithms provide us with a way to work with exponential functions. Some of the calculus operations we will encounter will have logarithms as a basis.

An exponential function is of the form: $y = a^x$, where x is a variable exponent and a is a numerical base greater than zero. We have developed the logarithmic form of an exponential function:

$$y = \log_a x$$

where a is known as the base $(a > 0)$ and y is the exponent. Our first task is to learn the procedure of symbol manipulation that allows us to convert from **exponential form** to **logarithmic form** and vice versa.

The following figure demonstrates the symbol placement needed to convert from one form to another.

$$25 = 5^2 \longrightarrow \log_5 25 = 2$$

Figure 1.6 Converting Between Forms

The base 5 is moved below and in front of the number 25, while the exponent 2 fills the location where the 5 was previously. The abbreviation "log" is added.

$$\log_2 8 = 3 \longrightarrow 8 = 2^3$$

Figure 1.7 Converting Between Forms

The base, 2, is moved to a base location below the exponent, 3. The abbreviation "log" is removed. The exponent, 3, is moved "up" with respect to the base number, 2.

EXAMPLE 1.18

Convert the following logarithms to exponential form.

a) $\log_2 x = 5$

b) $\log_x x = 1$

c) $\log_b w = g$

d) $\log_{1/2}\left(\frac{1}{8}\right) = 3$

SOLUTION 1.18

a) $\log_2 x = 5$ — Copy the problem.

$x = 2^5$ — Note that 2 is the base and that 5 is the exponent. Remove the abbreviation "log."

b) $\log_x x = 1$ — Copy the problem.

$x = x^1$ — Note that x is the base. Remove abbreviation "log."

$x = x$ — The result is the valid statement that $x = x$.

c) $\log_b w = g$ — Copy the problem.

$w = b^g$ — Note that b is the base and that g is the exponent. The abbreviation "log" is deleted.

d) $\log_{1/2}\left(\frac{1}{8}\right) = 3$ — Copy the problem.

$\frac{1}{8} = \left(\frac{1}{2}\right)^3$ — Note that 3 is the exponent and that 1/2 is the base. The abbreviation "log" is deleted. The parentheses on the 1/2 were added for clarity.

EXAMPLE 1.19

Convert the following expressions to logarithmic form:

a) $6 = x^2$

b) $1 = z^0$

c) $16 = 4^2$

d) $y = b^x$

SOLUTION 1.19

a) $6 = x^2$ — Copy the problem.

The base, x, is placed below the 6. The 2 is lowered.

$\log_x 6 = 2$	The abbreviation "log" is added.
b) $1 = z^0$	Copy the problem.
$\log_z 1 = 0$	The base, z, is placed below 1. The 0 exponent is lowered. The abbreviation "log" is added.
c) $16 = 4^2$	Copy the problem.
$\log_4 16 = 2$	The base, 4, is placed below 16. The 2 exponent is lowered. The abbreviation "log" is added.
d) $y = b^x$	Copy the problem.
$\log_b y = x$	The base, b, is placed below y. The x exponent is lowered. The abbreviation "log" is added.

Special Logarithmic Forms

Two particular bases, base 10 and base e, are used so often that they have special nomenclature of their own:

long form:	$\log_{10} x = y$	$\log_e x = y$
short form:	$\log x = y$	$\ln x = y$

Note that $\log_{10}$ can be expressed by simply writing "log." The symbols $\log_e$ are abbreviated "ln." The simpler forms (especially ln) are used in various sections of this text. (You might recal that e is the "natural number," or approximately 2.71828. . . .)

Rules of Logarithms

There are five useful rules for working with logarithms. These rules are listed below. We will explore operations with these rules.

I. $\log_a a = 1; \quad a > 0$

II. $\log_a 1 = 0; \quad a > 0$

III. $\log_a (x \cdot y) = \log_a x + \log_a y;$

IV. $\log_a \left(\dfrac{x}{y}\right) = \log_a x - \log_a y;$

V. $\log_a x^n = n\log_a x$.

These rules help us (a) expand a single logarithm into a series of logarithms, (b) combine a series of logarithms into a single logarithm, and (c) solve logarithmic equations.

EXAMPLE 1.20

Use the rules of logarithms to expand each of the following into a series of logarithms.

a) $\log 2y$ (base 10)

b) $\log_2\left(\frac{x}{3w}\right)$

c) $\log_4\left(\frac{x^2}{4y^3}\right)$

d) $\ln\frac{\sqrt{x}}{(x+1)}$ (base e)

SOLUTION 1.20

a) $\log 2y$ (base 10) — Copy the given logarithm.

$\log(2 \cdot y) = \log 2 + \log y$ — Note that 2 and y are factors and that these factors can be separated by using Rule III of logarithms.

b) $\log_2\left(\frac{x}{3w}\right)$ — Copy the given logarithm.

$\log_2\left(\frac{x}{3w}\right) = \log_2 x - \log_2 3w$ — Rule IV can be used to separate the fraction into the logarithm of the numerator minus the logarithm of the denominator.

$\log_2 x - (\log_2 3 + \log_2 w)$ — Rule III allows us to separate the factors of 3 and w. Parentheses had to be added because the negative sign belongs to both $\log_2 3$ and $\log_2 w$.

$\log_2\left(\frac{x}{3w}\right) = \log_2 x - \log_2 3 - \log_2 w$	Distribute the minus sign for the final answer.
c) $\log_4\left(\frac{x^2}{4y^3}\right)$	Copy the given logarithm.
$\log_4 x^2 - \log_4 4y^3$	The denominator of the fraction is subtracted from the numerator (use Rule IV).
$\log_4 x^2 - (\log_4 4 + \log_4 y^3)$	The factors 4 and y^3 are separated according to Rule III. (Remember parentheses).
$\log_4 x^2 - \log_4 4 - \log_4 y^3$	Distribute the minus sign.
$2\log_4 x - \log_4 4 - 3\log_4 y$	Use Rule V to move exponents.
$2\log_4 x - 1 - 3\log_4 y$	Use Rule I to write $\log_4 4 = 1$.
d) $\ln \frac{\sqrt{x}}{(x+1)}$ (base e)	Copy the logarithm.
$\ln \frac{x^{1/2}}{x+1}$	Rewrite the radical expression.
$\ln x^{1/2} - \ln(x+1)$	Subtract the ln of the denominator from the ln of the numerator according to Rule IV. The parentheses were added to $x+1$ for clarity.
$\frac{1}{2}\ln x - \ln(x+1)$	Use Rule V to move exponent.
$\frac{1}{2}\ln x - \ln(x+1)$	The final result does **not** separate $x+1$. The rules of logarithms are used to separate products and quotients, **not** sums and differences.

Combining as a Single Logarithm

Just as logarithmic expressions can be expressed as a sum or difference of logarithms, the sum or difference of logarithms can be combined to form a single logarithm.

EXAMPLE 1.21

Combine the following series of logarithms into a single logarithmic expression.

a) $\log_2 3 + \log_2 y + \log_2 x$

b) $2\log_7 x + 3\log_7 y - \log_7 z$

c) $\frac{1}{2}\log x - 3\log(x+1) - \log y - 2\log w$

d) $3 - 2\log_x y$

SOLUTION 1.21

a) $\log_2 3 + \log_2 y + \log_2 x$ — Copy the problem.

$\log_2 3 + \log_2 y + \log_2 x$

$= \log_2 3 \cdot y \cdot x$ — Each logarithm has the same base (2). We can work with Rule III to combine the logarithms.

$\log_2 3yx$ — The numbers and letters are combined as factors.

b) $2\log_7 x + 3\log_7 y - \log_7 z$ — Copy the problem.

$2\log_7 x + 3\log_7 y - \log_7 z$

$= \log_7 x^2 + \log_7 y^3 - \log_7 z$ — Use Rule V to move the coefficients of the logarithms to the exponent position.

$+\log_7 x^2 + \log_7 y^3 - \log_7 z$ — The coefficients have been moved. Note the sign (or assumed sign) preceeding each logarithm.

$\log_7 \frac{\quad}{\quad}$ — If both + and – signs are present, write the basic logarithmic symbol and the base. Follow this with a fraction bar.

$\log_7\left(\frac{x^2 \cdot y^2}{z}\right)$

Place the terms following a positive $\log_7$ above the fraction bar. Those terms following a negative $\log_7$ are placed below.

$\log_7\left(\frac{x^2 y^2}{z}\right)$

Note that x^2 and y^3 are **multiplied** as factors.

c) $\frac{1}{2}\log x - 3\log(x+1) - \log y - 2\log w$

Copy the problem. Since no base is shown, the base is assumed to be 10.

$\log x^{1/2} - \log(x+1)^3 - \log y - \log w^2$

$+\log x^{1/2} - \log(x+1)^3 - \log y - \log w^2$

Use Rule V to move the logarithm coefficients to the exponent position.

Note the signs of each logarithm.

$\log \frac{\quad}{\quad}$

Since both + and – signs are present, write a single logarithmic symbol followed by a fraction bar.

$\log \frac{x^{1/2}}{(x+1)^3 \cdot y \cdot w^2}$

Place all **positive** terms **above** the fraction bar and all **negative** terms **below** the fraction bar.

$\log \frac{\sqrt{x}}{(x+1)^3 y w^2}$

Rewrite $x^{1/2}$ as a radical. Note all terms are factors.

d) $3 - 2\log_x y$

Copy the problem.

The single number 3 cannot be combined with the other logarithm because it is **not** in logarithmic **form**.

$3 \cdot 1 - 2\log_x y$

We must remember that $3 = 3 \cdot 1$, and that $\log_x x = 1$ (Rule I).

$3 \cdot \log_x x - 2\log_x y$ — $\log_x x$ was used because the other logarithm was $\log_x$.

$\log_x x^3 - \log_x y^2$ — Use Rule V to move the coefficients of the logarithms to the exponent position.

$+\log_x x^3 - \log_x y^2$ — Note the signs.

$\log_x \text{———}$ — Since both + and – signs are present, write a single logarithm symbol with base of x followed by a fraction bar.

$\log_x \left(\frac{x^3}{y^2}\right)$ — Place x^3 above and y^2 below.

Solving Logarithmic Equations

Equations containing logarithms are often solved by combining logarithmic expressions (if necessary) and then rewriting in exponential form. A solution to a logarithmic equation may not be apparent until the equation is rewritten in exponential form and then solved by algebraic means. The following exercises provide a summary of problems typically encountered.

EXAMPLE 1.22

Solve for the unknown, x.

a) $\log_2 x = 3$

b) $\log_x 16 = 2$

c) $\log_3 27 = x$

d) $\log_2 x + \log_2 5 = 4$

e) $\log_5 (x-1) = 2$

f) $\log_3 x - \log_3 (x+1) = 2$

SOLUTION 1.22

a) $\log_2 x = 3$ — Copy the problem.

$\log_2 x = 3 \rightarrow x = 2^3$ — Rewrite in exponential form.

$x = 8$ — $2^3 = 8$, the solution.

b) $\log_x 16 = 2$	Copy the problem.
$16 = x^2$	Rewrite in exponential form.
$\sqrt{16} = \sqrt{x^2}$ $4 = x$	Take the square root of both sides. The positive root is the solution. (Negative numbers cannot be bases).
c) $\log_3 27 = x$	Copy the problem.
$27 = 3^x$	Rewrite in exponential form.
$27 = 3^3 = 3^x$ same base	Problems with unknown exponents are not easily solved unless, as in this case, we can write both sides of the equation with the same base.
$3^3 = 3^x$ $3 = x$ $x = 3$	Since the base for both sides of the equation is the same, and the expressions are equal, the exponents must be equal. This provides the answer.
d) $\log_2 x + \log_2 5 = 4$	Copy the problem.
$\log_2 x \cdot 5 = 4$	Use Rule III to combine the logarithmic expression.
$\log_2 5x = 4$	Rewrite in exponential form.
$5x = 2^4$	
$5x = 16$	$2^4 = 16$. Use algebra to solve for x.
$\frac{5x}{5} = \frac{16}{5}$	
$x = \frac{16}{5}$	

e) $\log_5(x-1) = 2$	Copy the problem.
$(x-1) = 5^2$	Rewrite in exponential form.
$x-1 = 25$	
$\begin{array}{rcl} x-1 &=& 25 \\ +1 && +1 \\ \hline x &=& 26 \end{array}$	Complete the solution through the use of algebra.
f) $\log_3 x - \log_3(x+1) = 2$	Copy the problem.
$\log_3\left(\dfrac{x}{x+1}\right) = 2$	Combine the logarithmic expressions using Rule IV.
$\dfrac{x}{x+1} = 3^2 = 9$	Rewrite in exponential form.
$\dfrac{x+1}{1} \cdot \dfrac{x}{x+1} = 9 \cdot (x+1)$	Multiply both sides by $x+1$.
$x = 9(x+1)$	Distribute the 9.
$\begin{array}{rcl} x &=& 9x+9 \\ -9x && -9x \\ \hline -8x &=& 9 \end{array}$	Complete the solution.
$\dfrac{-8x}{-8} = \dfrac{9}{-8}$ $x = -\dfrac{9}{8}$ No solution	There is no solution to this problem since we cannot have a negative number following a logarithm or as a logarithmic base.

Practice Exercises

1. Use the formulas for the sum and difference of two cubes to factor:

   a) $8x^3 - 1$

   b) $a^3 + 125b^3$

   c) $x^6 + y^6$

2. Factor the following using radical expressions.

   a) $y^2 - 6$

   b) $x^2 - 5$

   c) $x - 7$

3. Factor and simplify.

   a) $3(x-2)^2(x+3)^4 + 6(x-2)^3(x+3)^3$

   b) $2x(x-7)^7(x+8)^3 + 3x(x-7)^6(x+8)^4$

4. Convert the following radical expressions to terms with rational exponents.

   a) $\sqrt[3]{x^8}$

   b) $\sqrt[7]{x^{15}}$

   c) $\sqrt{x^5}$

5. Find the following products involving the multiplication of the expression using rational exponents.

   a) $(x^{1/4})(x^{1/2})$

   b) $x^{2/3}(x^{1/2} + x^{1/4})$

   c) $(x^{1/2} - x)(x + x^{1/3})$

6. Simplify the following algebraic expressions using subtraction of rational exponents.

   a) $\dfrac{x^4}{x^7}$

   b) $\dfrac{x^{3/5}}{\sqrt{x}}$

   c) $\dfrac{x^3 - 2x + 1}{x^{1/3}}$

7. Evaluate each function at the given domain element. Write the ordered pair for each problem.

	Function	Domain Element
a)	$f(x) = 6x + 2$	3
b)	$f(x) = \sqrt[3]{x}$	8
c)	$f(x) = 2x^3 + 3x + 1$	2
d)	$f(x) = 3x^2 + 7x$	$-s$

8. Use the difference quotient to evaluate the following functions.

   a) $f(x) = x - 3$

   b) $f(x) = \dfrac{1}{x-1}$

   c) $f(x) = x^2 + 3x + 2$

9. Find the slope of the line passing through

   a) $(-2, 4)$ and $(5, 3)$

   b) $(0, 3)$ and $(6, -1)$

10. Find the equation of

    a) a line with slope of $\frac{2}{3}$ and passing through $(1, -1)$.

    b) a line that connects $(5, -4)$ and $(3, 7)$.

    c) a vertical line passing through $(6, 3)$.

11. Solve the following inequalities.

    a) $2x - 1 > 11$

    b) $3 - 7x < 5$

    c) $4 - \frac{x}{2} > 6$

12. Find the solution to the following inequalities.

    a) $-1 < x - 6 < 4$

    b) $6 < 2 - 3x < 10$

13. Solve the following absolute value equations.

    a) $|x - 5| = 4$

    b) $\left|2 - \frac{x}{2}\right| = 5$

14. Solve the following absolute value inequalities.

    a) $|2 - x| > 3$

    b) $|3x - 3| < 1$

15. Solve the following equations and show their solution(s) on the number line.

    a) $2x - 1 = 3$

    b) $|x - 6| = 4$

16. Use interval notation to describe the solutions of the following inequalities. Show the graph, also.

    a) $|x - 7| < 3$

    b) $|2 - 6x| > 5$

    c) $7 - \frac{x}{3} < 1$

17. Find the restrictions on the domain of the following.

a) $f(x) = \frac{7}{x+1}$

b) $f(x) = \sqrt{12-x}$

c) $f(x) = \frac{\sqrt{x-1}}{3x+7}$

d) $f(x) = \frac{x}{x^2+1}$

e) $f(x) = \frac{4x}{\sqrt{x-2}}$

18. Convert the following equations to logarithmic form.

a) $2 = 4^{1/2}$

b) $x^2 = 8$

c) $a^w = z$

19. Convert the following logarithms to exponential form.

a) $\log_4 16 = 2$

b) $\ln x = 17$

c) $\log 1000 = 3$

20. Use the rules of logarithms to expand the following.

a) $\log_2 \frac{3x^2}{y}$

b) $\log \frac{(x+1)^2}{\sqrt{(x+y)}}$

c) $\ln \frac{4xy}{xy-16}$

21. Use the rules of logarithms to combine the logarithms.

a) $\log_3 5 + \log_3 x - 2\log_3 y$

b) $\log(x+1) - \frac{1}{2}\log(x-5) - \frac{1}{2}\log(xy-1)$

c) $\ln 6 + \ln x - \frac{1}{2}\ln y - \frac{3}{2}\ln(x^2-2)$

22. Solve the following logarithmic equations.

a) $\log_3 x = 3$

b) $\log_x 2 = 2$

c) $\log_2 8 = x$

d) $\log_3 x - \log_3 12 = 1$

e) $\log_2 x + \log_2(x+1) = \log_2 12$

Answers

1. a) $(2x-1)(4x^2+2x+1)$

   b) $(a+5b)(a^2-5ab+25b^2)$

   c) $(x^2+y^2)(x^4-x^2y^2+y^4)$

2. a) $(y-\sqrt{6})\,(y+\sqrt{6})$

   b) $(x-\sqrt{5})\,(x+\sqrt{5})$

   c) $(\sqrt{x}-\sqrt{7})\,(\sqrt{x}+\sqrt{7})$

3. a) $3\,(3x-1)\,(x-2)^2\,(x+3)^3$

   b) $5x\,(x+2)\,(x-7)^6\,(x+8)^3$

4. a) $x^{8/3}$

   b) $x^{15/7}$

   c) $x^{5/2}$

5. a) $x^{3/4}$

   b) $x^{7/6}+x^{11/12}$

   c) $x^{3/2}+x^{5/6}-x^2-x^{4/3}$

6. a) x^{-3}

   b) $x^{1/10}$

   c) $x^{8/3}-2x^{2/3}+x^{-1/3}$

7. a) $f(3)\;=\;20\;\;(3,20)$

   b) $f(8)\;=\;2\;\;(8,2)$

   c) $f(2)\;=\;23\;\;(2,23)$

   d) $f(-s)\;=\;3s^2-7s$

8. a) 1

   b) $\dfrac{-1}{(x-1)\,(x+h-1)}$

   c) $2x+3+h$

9. a) $-1/7$

   b) $-2/3$

10. a) $2x-3y-5=0$

    b) $11x+2y-47=0$

    c) $x=6$

11. a) $(6,\infty)$

    b) $(-2/7,\infty)$

    c) $(-\infty,-4)$

12. a) $(5, 10)$

    b) $(-8/3,-4/3)$

13. a) $x=1, x=9$

    b) $x=-6, x=14$

14. a) $(-\infty,-1)\cup(5,\infty)$

    b) $(2/3, 4/3)$

15. a) $x=2$

    b) $x=2, x=10$

16. a) $(4, 10)$

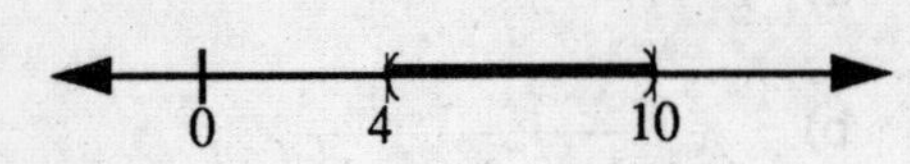

b) $\left(-\infty, -\frac{1}{2}\right) \cup \left(\frac{7}{6}, \infty\right)$

-1/2 0 7/6

c) $(18, \infty)$

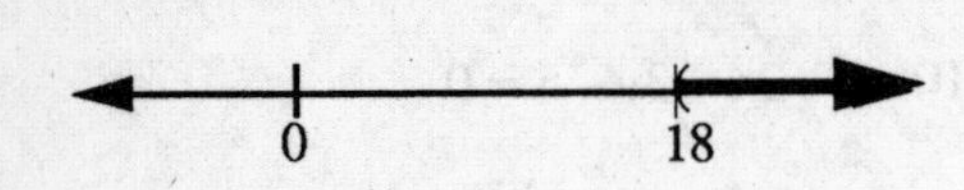

17. a) restriction at $x = -1$

b) restriction for all $x > 12$

c) restriction for all $x < 1$

d) no restrictions

e) restriction for all $x \leq 2$

18. a) $\log_4 2 = 1/2$

b) $\log_x 8 = 2$

c) $\log_a z = w$

19. a) $4^2 = 16$

b) $e^{17} = x$

c) $10^3 = 1000$

20. a) $\log_2 3 + 2\log_2 x - \log_2 y$

b) $2\log(x+1) - \frac{1}{2}\log(x+y)$

c) $\ln 4 + \ln x + \ln y - \ln(xy - 16)$

21. a) $\log_3 \frac{5x}{y^2}$

b) $\log \frac{x+1}{\sqrt{(x-5)(xy-1)}}$

c) $\ln \frac{6x}{\sqrt{y(x^2-2)^3}}$

22. a) $x = 27$

b) $x = \sqrt{2}$

c) $x = 3$

d) $x = 36$

e) $x = 3$ ($x = -4$ is undefined)

2

Limits and Continuity

2.1 INTRODUCTION TO LIMITS

The basic foundations of calculus are built upon the concept of limits. This concept is, perhaps, one of the most intriguing ideas found in a calculus course. As we proceed through this chapter, we will focus on the following topics:

a) The basic concepts of limits
b) Graphic interpretation of limits
c) Algebraic techniques used in determining limits
d) Finding limits *of* infinity
e) Finding limits *at* infinity
f) Checking for continuity

The Concept of Limits

A graph is necessary to help clarify how limits are viewed in calculus. Consider a function labeled by the general function notation of $f(x)$. It is graphed in the first quadrant of the rectangular coordinate system.

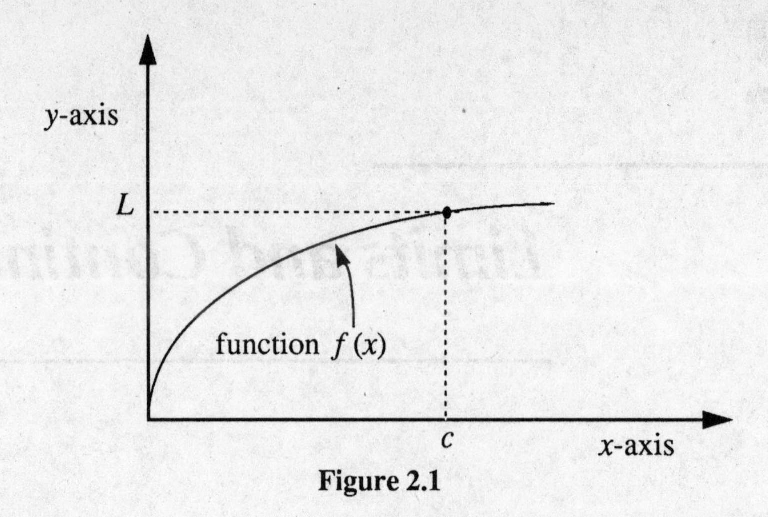

Figure 2.1

The location on the x-axis labeled c is matched with the function to determine the limit value, L, which is read from the y-axis. The symbolic representation for a limit under these conditions is:

2.1 Limit Formula

$$\lim_{x \to c} f(x) = L$$

where:

lim is read "the limit"
$x \to c$ means "as x approaches c"
$f(x)$ is any given function ($x^2 - 1$, for example)
L is the limit value

If numbers are substituted for c and L and a function substituted for $f(x)$, we could, for example, have

$$\lim_{x \to 2} (x^2 - 1) = 3$$

The example would read "the limit, as x approaches 2, of $x^2 - 1$ is 3." Note that when 2 is substituted into $x^2 - 1$, we have $2^2 - 1 = 4 - 1 = 3$ (the limit). It is through the substitution of the c-value into a given function that a limit, L, will be found. Unfortunately, this is not always simple.

Other aspects of the determination of limits can be shown by looking at the graph of

$$f(x) = \frac{x^3 - 3x^2 + x - 3}{x - 3}$$

Figure 2.2 shows this graph with an open circle at the (x, y) location of (3, 10).

This open circle indicates that this function does not exist at (3, 10), because a substitution of $x = 3$ in this function causes division by zero.

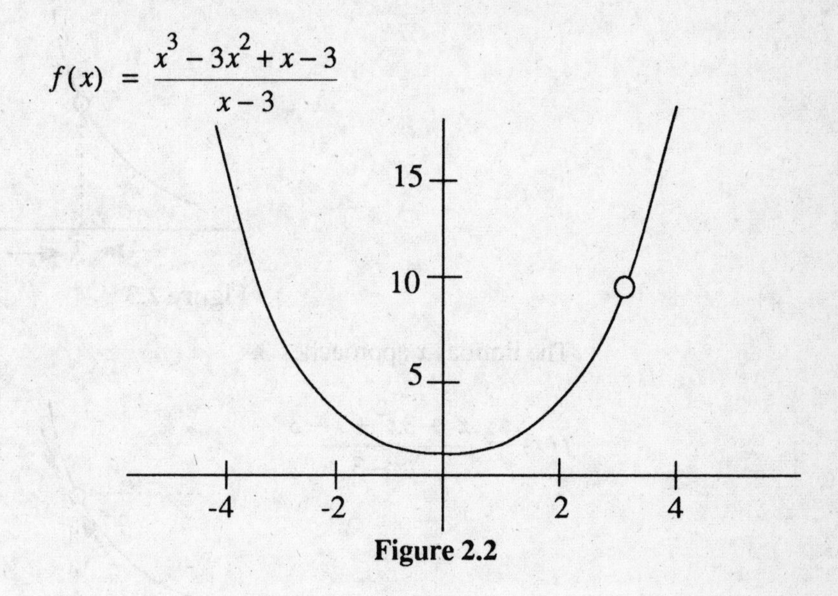

Figure 2.2

This graph shows that if this function were to exist at $x = 3$, then a function value of 10 would be the best choice. We can state this in symbolic terms as

$$\lim_{x \to 3} \frac{x^3 - 3x^2 + x - 3}{x - 3} = 10$$

Even though we cannot actually let $x = 3$, we can choose values very near 3 to get an idea of what value the function approaches as x approaches 3. Let's substitute $x = 2.99$ and $x = 3.01$ in the function

$$f(x) = \frac{x^3 - 3x^2 + x - 3}{x - 3}$$

$$f(2.99) = \frac{(2.99)^3 - 3(2.99)^2 + (2.99) - 3}{2.99 - 3} = 9.99$$

$$f(3.01) = \frac{(3.01)^3 - 3(3.01)^2 + (3.01) - 3}{3.01 - 3} = 10.01$$

As you can see, the values are very close to 10.

Figures 2.3, 2.4, and 2.5 show the observations we must make as we examine each limit.

The limit as x approaches 3 $\lim\limits_{x \to 3}$

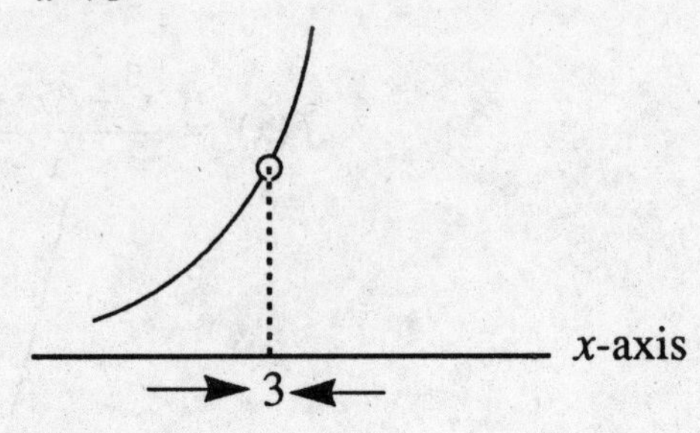

Figure 2.3

The limit as x approaches 3.

$$f(x) = \frac{x^3 - 3x^2 + x - 3}{x - 3}$$

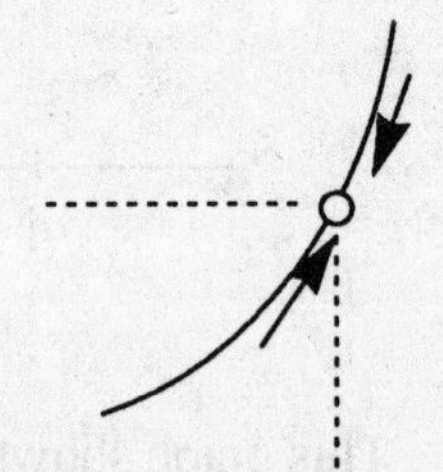

Figure 2.4

The corresponding y-axis value is equal to 10 ($L = 10$).

$$\lim_{x \to 3} f(x) = 10$$

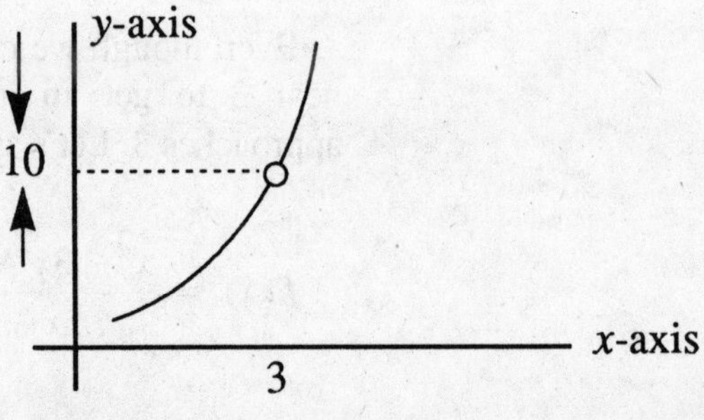

y-axis
10
x-axis
3

Figure 2.5

Figures 2.3, 2.4, and 2.5 point out the 3-part nature of searching for a limit. It is very important that you constantly focus on each of the three locations on the graph that pertain to the limit:

1. The designated point on the x-axis.
2. The segment of the function vertically above (or below) this x-axis location.
3. The number on the y-axis that is directly across from the important segment of the function.

The next section will demonstrate special aspects of determining limits and function values from visual inspection of graphs.

2.2 GRAPHIC INTERPRETATION OF LIMITS

Many texts present the graph of a unique function and ask you to determine values at various points along the x-axis. This section will highlight the more important aspects of this technique.

Table 2.1 summarizes the common formats found in graphs and their meaning. Each format is examined for a function value and the limit value. A third aspect (continuity) will be discussed later in the chapter.

Type of Format	Visual Display	Meaning
solid curve	c	limit exists at c, function exists at c
solid curve interupted by a solid dot	c	limit exists at c, function exists at c

Table 2.1 Graphic Representation of Limits

Type of Format	Visual Display	Meaning
solid curve interrupted by an open circle	c	limit exists at c, function does not exist at c
solid curve interrupted by an open circle. Solid dot above or below open circle.	c	limit exists at c at level of open circle, function exists at level of solid dot.
two solid curves terminating at different levels above c	c	limit does not exist. (We do not have the same y-value at the left and right of c.) Function exists at solid dot location.
vertical asymptote occurs at c	c	function does not exist at c. Limit does not exist at c.

Table 2.1 Graphic Representation of Limits (con't.)

We must remember that the value of c on the x-axis is approached from both positive and negative directions simultaneously. Therefore, it is important to trace the function from *both* directions as we narrow our focus on the area directly above point c on the x-axis.

An exception to this process is when we use one-sided limits. One-sided limits will be addressed at the end of this chapter.

One final point before we begin an exercise on determining limit values and function values. Students often have trouble with the difference between *approaching* a limit and actually being *at* a function value.

Consider the following diagram with the open circle on a function "magnified."

Magnification of a limit

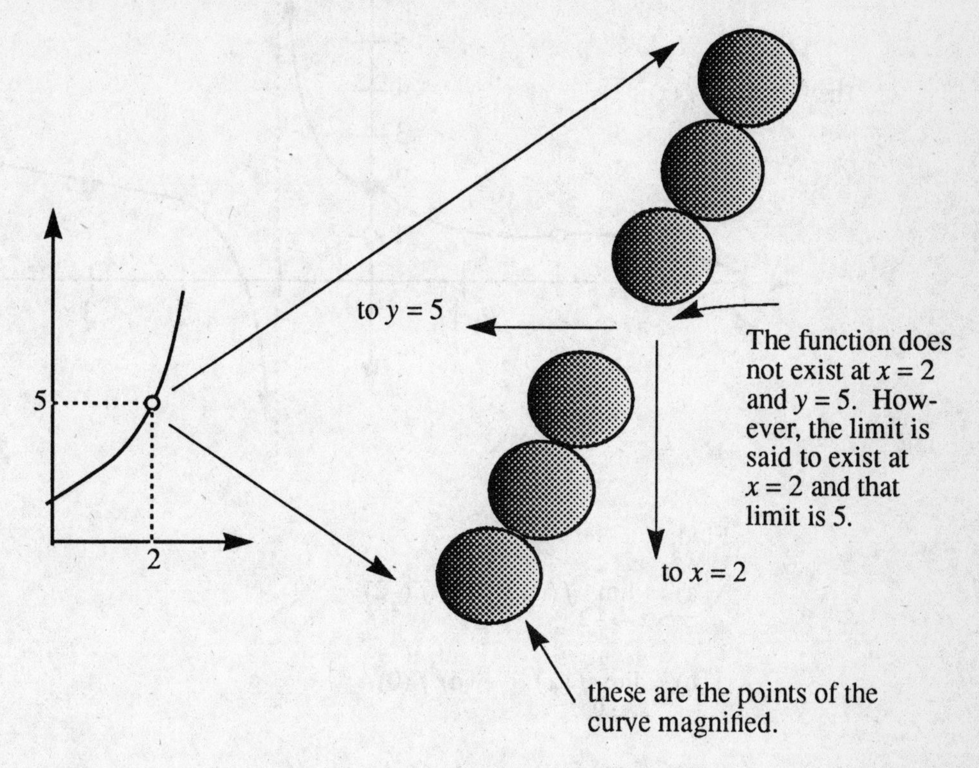

Figure 2.6

Study Figure 2.6. We know that the function does *not* exist at $x = 2$. However, we say that the limit is 5 when x approaches 2.

EXAMPLE 2.1

Using the ideas presented in Table 2.1 determine the values for the given functions in the graph on the following page.

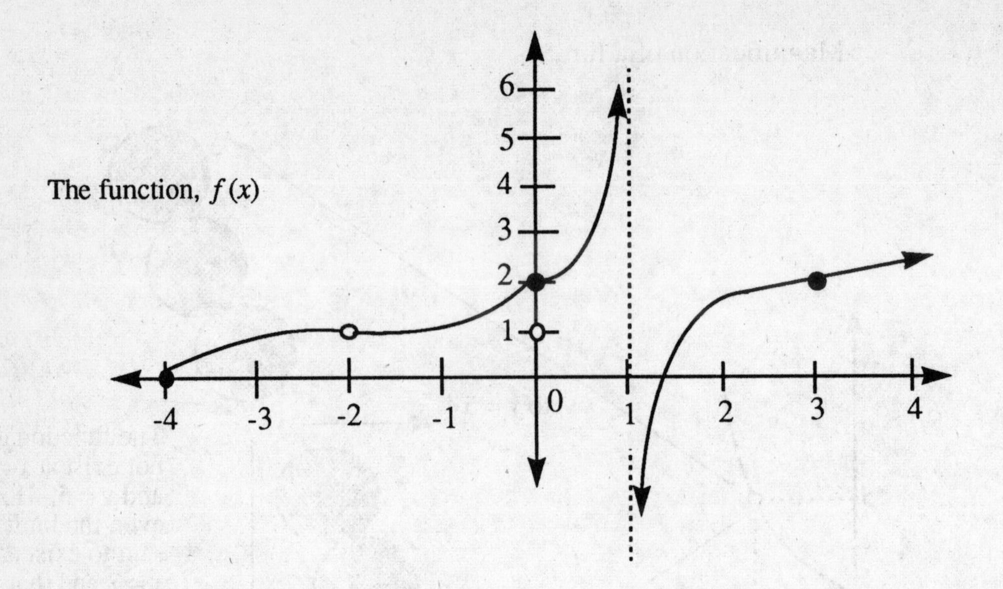

Find:

a) $\lim_{x \to -2} f(x)$ b) $f(-2)$

c) $\lim_{x \to 0} f(x)$ d) $f(0)$

e) $\lim_{x \to 1} f(x)$ f) $f(1)$

g) $\lim_{x \to 2} f(x)$ h) $f(2)$

i) $\lim_{x \to 3} f(x)$ j) $f(3)$

k) $\lim_{x \to -4} f(x)$ l) $f(-4)$

SOLUTION 2.1

a) $\lim_{x \to -2} f(x) = 1$

Refer to the graph at $x = -2$. As we approach -2 on the x-axis from both directions, it appears as though we are heading toward the y-level of 1 on the function.

b) $f(-2) = 2$

Refer to the graph at $x = -2$. Vertically above $x = -2$ is a solid dot even with 2 on the y-axis. This is the function value.

c) $\lim_{x \to 0} f(x)$ does not exist (DNE)

As we approach 0 on the x-axis from both directions, it appears that we are at the $y = 1$ level coming from the right and at the $y = 2$ level coming from the left. We must approach the same y-value as we come from both directions for the limit to exist.

d) $f(0) = 2$

Refer to graph at $x = 0$. The solid dot is at 2 on the y-axis.

e) $\lim_{x \to 1} f(x)$ does not exist (DNE)

Refer to graph at $x = 1$. Limit does not exist since right side of function approaches negative direction and left side approaches positive direction.

f) $f(1)$ does not exist (DNE)

Refer to graph at $x = 1$. A function does not exist at a vertical asymptote.

g) $\lim_{x \to 2} f(x) = 1$

Refer to graph at $x = 2$. It appears as though we are approaching the y-value of 1 when we move along $f(x)$ from both directions.

h) $f(2) = 1$

Refer to graph at $x = 2$. The function exists at $y = 1$ when $x = 2$. No solid dot is necessary.

i) $\lim_{x \to 3} f(x) = 2$	Refer to the graph at $x = 3$. As we travel along the function approaching $x = 3$ from both directions, it appears we are at the level where $y = 2$.
j) $f(3) = 2$	Refer to the graph at $x = 3$. The y-value of the function at $x = 3$ is $y = 2$.
k) $\lim_{x \to -4} f(x)$ does not exist (DNE)	Refer to the graph at $x = -4$. We cannot approach $x = -4$ from the left at this location because the function does not exist to the left of $x = -4$.
l) $f(-4) = 0.5$	Refer to the graph at $x = -4$. The function exists at about the $y = 0.5$ level at $x = -4$.

2.3 THE ALGEBRA USED IN DETERMINING LIMITS

While the visual inspection of a graph is one way to determine limits, another process involves the use of algebraic technique. Initially, some functions do not appear to have a limit at the given x-value. It is in these cases that the use of the algebra processes demonstrated in chapter 1 *may* be useful. The word "may" is stressed here because some functions do not have limits at various x-values.

Table 2.2 shows the most common algebraic techniques used in determining limits.

Technique	**Example**
Direct substitution	$\lim_{x \to 2} (3x - 5) = 3(2) - 5 = 1$
Removing a common factor	$\frac{x^3 - 2x^2 + x}{x} = \frac{x(x^2 - 2x + 1)}{x}$ $= x^2 - 2x + 1$

Table 2.2 Algebraic Techniques Used in Determining Limits

Technique	Example
Factoring difference of two squares	$\frac{x^2-9}{x-3} = \frac{(x+3)\,(x-3)}{(x-3)}$ $= x+3$
Rationalizing the numerator	$\frac{x-\sqrt{3}}{x^2-3} \cdot \frac{x+\sqrt{3}}{x+\sqrt{3}}$ $= \frac{(x^2-3)}{(x^2-3)\,(x+\sqrt{3})} = \frac{1}{x+\sqrt{3}}$
Factoring a trinomial	$\frac{x^2+7x+10}{x+2}$ $= \frac{(x+5)\,(x+2)}{(x+2)} = x+5$

Table 2.2 Algebraic Techniques Used in Determining Limits (con't.)

EXAMPLE 2.2

Find the limit for the following functions.

a) $\lim_{x \to 2} (3x^2 - 7)$

b) $\lim_{x \to 0} \frac{x^2-x}{\sqrt{x}}$

c) $\lim_{x \to 5} \frac{x^2-25}{x-5}$

d) $\lim_{x \to 2} \frac{\sqrt{x}-\sqrt{2}}{x-2}$

e) $\lim_{x \to -3} \frac{2x^2+7x+3}{x+3}$

f) $\lim_{x \to 10} 5$

g) $\lim_{x \to 7} \frac{4}{7-x}$

SOLUTION 2.2

a) $\lim_{x \to 2} (3x^2 - 7)$ — Copy the function.

$3(2)^2 - 7 = 3(4) - 7 = 12 - 7 = 5$ — Use substitution.

$\lim_{x \to 2} (3x^2 - 7) = 5$ — The limit equals 5.

b) $\lim_{x \to 0} \frac{x^2 - x}{\sqrt{x}}$ — Copy the function.

$\frac{0^2 - 0}{\sqrt{0}} = \frac{0}{0}$ undefined — Substitution does not yield a solution.

$\lim_{x \to 0} \frac{x^2 - x}{\sqrt{x}} = \lim_{x \to 0} \frac{x^2 - x}{x^{1/2}}$ — The limit cannot be evaluated when the function is in this form. Remove common factor.

$= \lim_{x \to 0} \frac{x^{1/2} x^{3/2} - x^{1/2} x^{1/2}}{x^{1/2}}$

$= \lim_{x \to 0} \frac{x^{1/2} (x^{3/2} - x^{1/2})}{x^{1/2}}$ — Remove $x^{1/2}$ and cancel.

$= \lim_{x \to 0} (x^{3/2} - x^{1/2}) =$

$= 0^{3/2} - 0^{1/2} = 0$ — The limit is 0.

c) $\lim_{x \to 5} \frac{x^2 - 25}{x - 5}$ — Copy the function.

$\frac{5^2 - 25}{x - 5} = \frac{25 - 25}{5 - 5} = \frac{0}{0}$ undefined — The limit cannot be evaluated when the function is in this form.

$\lim_{x \to 5} \frac{x^2 - 25}{x - 5} = \lim_{x \to 5} \frac{(x - 5)(x + 5)}{(x - 5)}$ — Factor the numerator as a difference of two squares.

$\lim_{x \to 5} \frac{\cancel{(x - 5)}(x + 5)}{\cancel{(x - 5)}}$ — Cancel common factors.

$\lim_{x \to 5} (x + 5) = 5 + 5 = 10$ — The limit can be evaluated.

d) $\lim_{x \to 2} \frac{\sqrt{x} - \sqrt{2}}{x - 2}$ — Copy the function.

$\frac{\sqrt{2} - \sqrt{2}}{2 - 2} = \frac{0}{0}$, undefined — The limit cannot be evaluated when the function is in this form.

$= \lim_{x \to 2} \frac{\sqrt{x} - \sqrt{2}}{x - 2} \cdot \frac{\sqrt{x} + \sqrt{2}}{\sqrt{x} + \sqrt{2}}$ — Rationalize the numerator. The denominator must also be multiplied, but do not distribute in the denominator.

$= \lim_{x \to 2} \frac{(x - 2)}{(x - 2)(\sqrt{x} + \sqrt{2})}$

$= \lim_{x \to 2} \frac{\cancel{(x - 2)}}{\cancel{(x - 2)}(\sqrt{x} + \sqrt{2})}$ — Cancel the common factors.

$= \lim_{x \to 2} \frac{1}{\sqrt{x} + \sqrt{2}} = \frac{1}{\sqrt{2} + \sqrt{2}}$ — Evaluate the function.

$= \frac{1}{2\sqrt{2}}$ — The limit is now found.

e) $\lim_{x \to -3} \frac{2x^2 + 7x + 3}{x + 3}$ — Copy the function.

$\frac{2(-3)^2 + 7(-3) + 3}{-3 + 3} = \frac{18 - 21 + 3}{0}$ — Check to see if the limit exists at the given x-value.

$= \frac{21 - 21}{0} = \frac{0}{0}$ — The limit cannot be evaluated when the function is in this form.

$\lim_{x \to -3} \frac{2x^2 + 7x + 3}{x + 3}$

$= \lim_{x \to -3} \frac{(2x + 1)(x + 3)}{(x + 3)}$ — Factor the numerator.

$= \lim_{x \to -3} \frac{(2x + 1)\cancel{(x + 3)}}{\cancel{(x + 3)}}$ — Cancel common factors.

$= \lim_{x \to -3} (2x + 1) = 2(-3) + 1$ — The limit can now be found.

$= -6 + 1 = -5$

f) $\lim_{x \to 10} 5$

Copy the function.

$\lim_{x \to 10} 5 = 5$

There is no way to introduce x-values into this function. The limit for this function will always be 5, regardless of what value x approaches.

g) $\lim_{x \to 7} \frac{4}{7-x}$

Copy the function.

$\frac{4}{7-7} = \frac{4}{0}$ undefined

The function has no limit at this x-value.

There is no algebraic technique available to convert this function into a form to obtain a limit. Many students find it difficult to determine when algebra works and when it does not. The best solution is to practice with many problems.

2.4 LIMITS OF INFINITY

There are two types of limit problems that involve infinity. These two types are expressed, symbolically, below:

I. $\lim_{x \to c} f(x) = \infty$ Limit *of* infinity

II. $\lim_{x \to \infty} f(x) = L$ Limit *at* infinity

Limits of infinity are covered in this section.

The limit obtained as we approach a value on the x-axis has been discussed. It is an approach from both the left and the right as we near the designated value. In order to more easily understand limits *of* infinity, we will focus on one-sided limits. These limits are determined when we approach the x-value from only one direction. Formulas 2.2 and 2.3 show the symbols used to indicate one-sided limits.

2.2	$\lim_{x \to c^+} f(x) = L$	The limit as x approaches c from the positive (right) side.
2.3	$\lim_{x \to c^-} f(x) = L$	The limit as x approaches c from the negative (left) side.

The graph of the function $f(x) = 1/(x-1)$ shows the difference in one-sided limits as x approaches 1 on the x-axis. Figure 2.7 shows the graph of $f(x) = \dfrac{1}{x-1}$.

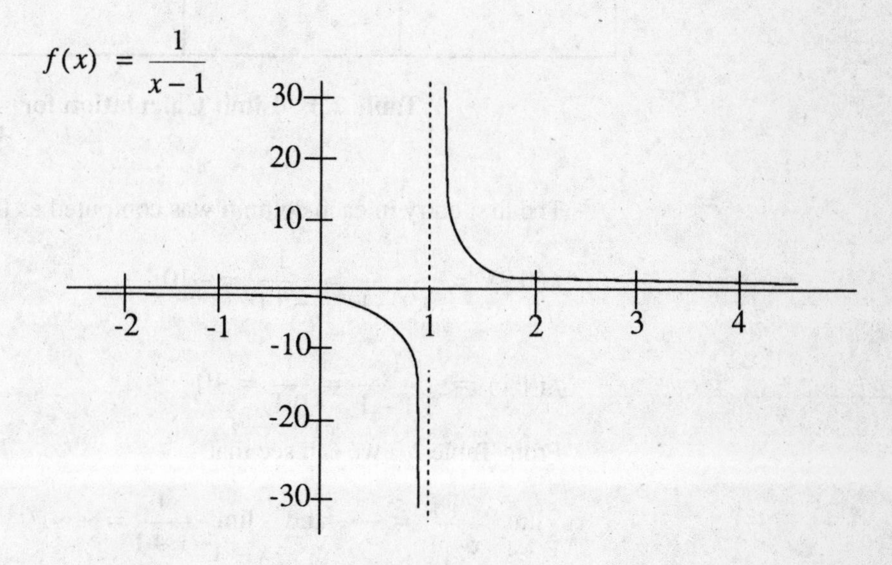

Figure 2.7

The graph shows that the function becomes a very large *negative* number as x nears 1 from the *left* side. It also shows that the function is a very large *positive* number as x approaches 1 from the right side. Table 2.3 shows the numerical values for the function as x approaches 1.

Approach Left		Approach Right	
x	$f(x)$	x	$f(x)$
0.9	–10	1.1	10
0.99	–100	1.01	100
0.999	–1000	1.001	1000
0.9999	–10000	1.0001	10000
•	•	•	•
•	•	•	•
	$-\infty$		∞

Table 2.3 Limit Calculation for $\dfrac{1}{x-1}$

The first entry in each column was computed as follows:

$$f(0.9) = \frac{1}{0.9-1} = \frac{1}{-0.1} = -10$$

$$f(1.1) = \frac{1}{1.1-1} = \frac{1}{0.1} = 10$$

From Table 2.3 we can see that

$$\lim_{x \to 1^-} \frac{1}{x-1} = -\infty \text{ and } \lim_{x \to 1^+} \frac{1}{x-1} = +\infty$$

While the one-sided limits (at 1) exist for $f(x) = 1/(x-1)$, the overall limit (approaching 1 from both sides) does *not* exist because the function does not move toward the same location on the graph. Another way to confirm a limit, as x approaches from both directions, is to check to see if the right-hand and left-hand limits are equal. If the one-sided limits are equal, the overall limit exists. For a given x-value, the overall limit does *not* exist when the one-sided limits are not equal.

The function $\dfrac{1}{x^2}$ has a limit as x approaches 0, since the one-sided limits are equal.

$$\lim_{x \to 0^-} \frac{1}{x^2} = \infty \qquad \lim_{x \to 0^+} \frac{1}{x^2} = \infty$$

Table 2.4 shows some of the calculations for these limits.

Approach Left		Approach Right	
x	$f(x)$	x	$f(x)$
–.1	100	.1	100
–.01	10000	.01	10000
–.001	1000000	.001	1000000
•	•	•	•
•	•	•	•
	∞		∞

Table 2.4 Calculation of Limit for $f(x) = 1/x^2$

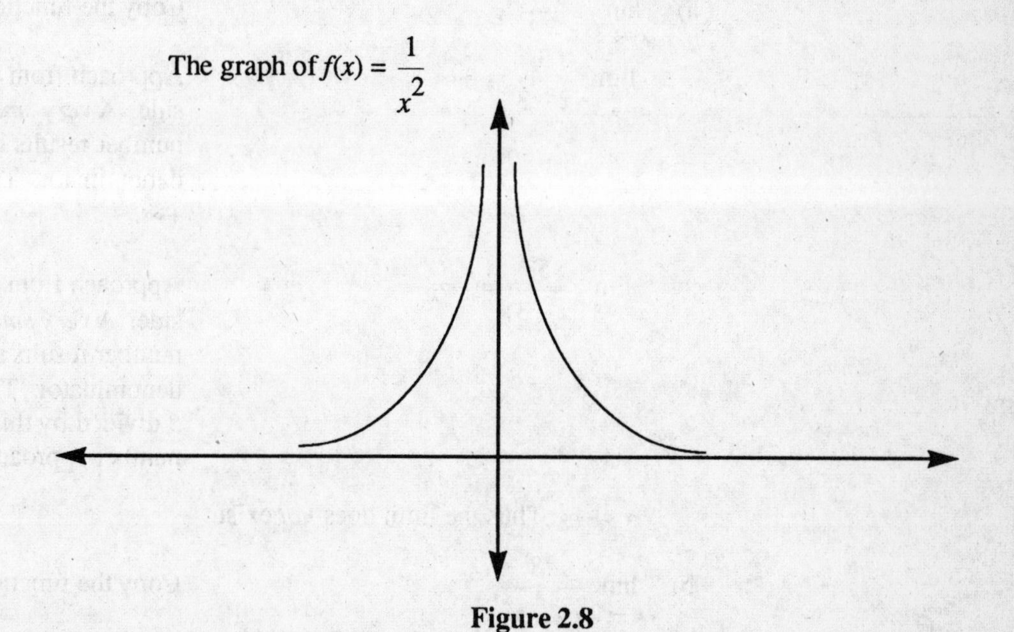

Figure 2.8

Figure 2.8 shows that both segments of the graph of the function $f(x) = 1/x^2$ travel up the y-axis (and, thus, toward $+\infty$).

While infinity is a number that can never be reached, it is suitable for a limit. Remember that a limit is approached, but never reached.

EXAMPLE 2.3

Determine the limit for the following functions, if possible. Use one-sided limits to test for a limit.

a) $\lim\limits_{x \to 3} \dfrac{5}{x-3}$

b) $\lim\limits_{x \to 2} \dfrac{8}{(x-2)^2}$

c) $\lim\limits_{x \to 5} \dfrac{1-x}{(x-5)^2}$

SOLUTION 2.3

a) $\lim\limits_{x \to 3} \dfrac{5}{x-3}$ — Copy the function.

$\lim\limits_{x \to 3^+} \dfrac{5}{x-3} = +\infty$ — Approach from the right (+) side. A *very small positive* number results in the denominator. The limit is $+\infty$.

$\lim\limits_{x \to 3^-} \dfrac{5}{x-3} = -\infty$ — Approach from the left (–) side. A *very small negative* number results in the denominator. The positive 5 divided by this negative number approaches $-\infty$.

$\infty \neq -\infty$ Thus the limit does *not* exist.

b) $\lim\limits_{x \to 2} \dfrac{8}{(x-2)^2}$ — Copy the function.

$$\lim_{x \to 2^+} \frac{8}{(x-2)^2} = +\infty$$

Approach from the right (+) side. The denominator becomes a very small positive number squared. A positive 8 divided by an increasingly small positive number approaches $+\infty$.

$$\lim_{x \to 2^-} \frac{8}{(x-2)^2} = -\infty$$

Approach from the left (–) side. The number becomes a very small negative number squared. The denominator is *positive* overall. A limit of $+\infty$ occurs.

The limit exists and equals ∞.

$$\lim_{x \to 2} \frac{8}{(x-2)^2} = \infty$$

c) $$\lim_{x \to 5} \frac{1-x}{(x-5)^2}$$

Copy the function.

$$\lim_{x \to 5^+} \frac{1-x}{(x-5)^2} = -\infty$$

Approach from the right (+) side. This gives a very small positive number squared in the denominator. $1 - x = 1 - 5 = -4$. -4 divided by an increasingly small positive number approaches $-\infty$.

$$\lim_{x \to 5^-} \frac{1-x}{(x-5)^2} = -\infty$$

Approach from the left (–) side. This gives a very small negative number squared in the denominator. The denominator is positive. -4 divided by a small positive number is a large negative number. The limit exists and equals $-\infty$.

2.5 LIMITS AT INFINITY

The second type of limit problem involving infinity occurs when x approaches either negative infinity or positive infinity.

There is a general rule involving a reciprocal power of x. The formula involved is as follows:

2.4 $\lim\limits_{x \to \infty} \dfrac{1}{x^p} = 0$ where p is any real number greater than zero.

This formula allows us to evaluate an expression like:

$$\lim_{x \to \infty} \frac{2x^2 - x}{3x^2 + 1}$$

by dividing each of the terms by the highest power of x (x^2 in this case).

The function now becomes:

$$\lim_{x \to \infty} \frac{\dfrac{2x^2}{x^2} - \dfrac{x}{x^2}}{\dfrac{3x^2}{x^2} + \dfrac{1}{x^2}} = \lim_{x \to \infty} \frac{2 - \dfrac{1}{x}}{3 + \dfrac{1}{x^2}}$$

The terms $1/x$ and $1/x^2$ are essentially zero when x is very large (divide 1 by successively larger numbers and see how the quotient becomes very small). The limit is now found.

$$\lim_{x \to \infty} \frac{2 - \dfrac{1}{x}}{3 + \dfrac{1}{x^2}} = \frac{2-0}{3+0} = \frac{2}{3}$$

The process of dividing by the largest power of x is very time-consuming. This process can be shortened by observing each of the three cases that occur with this type of problem.

Case	Highest Power	Limit	Example
I	In numerator	$\pm\infty$	$\lim_{x \to \infty} \frac{x^3 - 1}{x^2} = \infty$
II	Equal in both numerator and denominator	Ratio of coefficients of highest power	$\lim_{x \to \infty} \frac{3x^2 - 1}{7x^2 - x} = \frac{3}{7}$
III	In denominator	0	$\lim_{x \to \infty} \frac{5 - x}{x^3 + 2} = 0$

Table 2.5 Limits at Infinity for Rational Functions

EXAMPLE 2.4

Find the following limits *at* infinity.

a) $\lim_{x \to \infty} \frac{x^3 + 2x - 1}{x^2 + 8x}$

b) $\lim_{x \to \infty} \frac{x + 1}{x^3 + 2}$

c) $\lim_{x \to \infty} \frac{8x^2 + x}{11x^2 + 2}$

d) $\lim_{x \to -\infty} \frac{x^3 + x}{x^2 - 2x}$

SOLUTION 2.4

a) $\lim_{x \to \infty} \frac{x^3 + 2x - 1}{x^2 + 8x}$ Copy the function.

$\lim_{x \to \infty} \frac{x^3 + 2x - 1}{x^2 + 8x} = \infty$ Since the highest power is in the numerator, the limit is ∞.

b) $\lim_{x \to \infty} \frac{x+1}{x^3+2}$ — Copy the function.

$\lim_{x \to \infty} \frac{x+1}{x^3+2} = 0$ — The highest power is in the denominator. The limit is zero.

c) $\lim_{x \to \infty} \frac{8x^2+x}{11x^2+2}$ — Copy the function.

$\lim_{x \to \infty} \frac{8x^2+x}{11x^2+2} = \frac{8}{11}$ — The largest exponents are the same in the numerator and the denominator. The limit is the ratio of the coefficients.

d) $\lim_{x \to -\infty} \frac{x^3+x}{x^2-2x}$ — Copy the function.

$\lim_{x \to -\infty} \frac{x^3+x}{x^2-2x}$ — The largest exponent is in the numerator.

$\lim_{x \to -\infty} \frac{x^3+x}{x^2-2x} = -\infty$ — The value of x^3 is *negative* and the value of x^2 is *positive* as x approaches $-\infty$. The quotient is *negative*.

2.6 CONTINUITY

A function is continuous if it follows a smooth uninterrupted path. The function cannot have any breaks or "holes." There is a strict definition for continuity. A function is continuous at c if:

(I)	$\lim_{x \to c} f(x) = L$	(the limit exists)
(II)	$f(c)$ exists	(the function value at c exists)
(III)	$f(c) = L$	(the function value at c equals the limit)

EXAMPLE 2.5

The graph of function, $f(x)$, is shown below. Determine whether the function is continuous at each value.

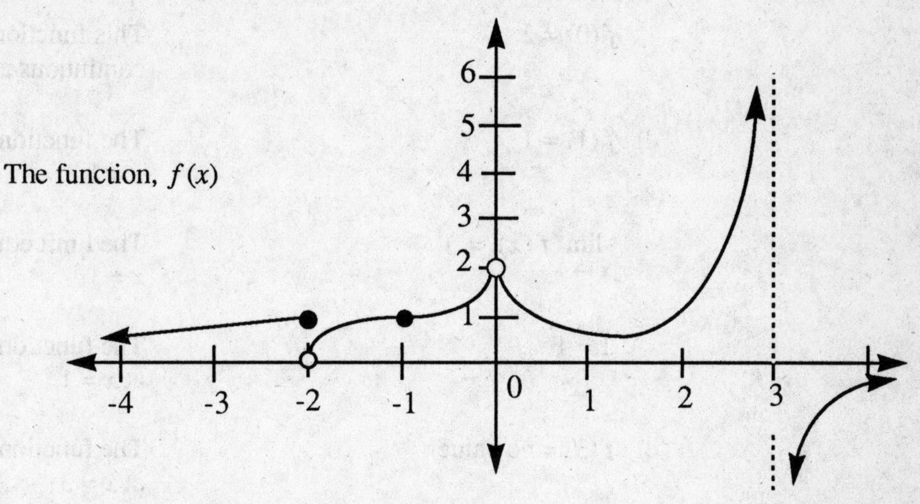

a) at $x = -2$

b) at $x = -1$

c) at $x = 0$

d) at $x = 1$

e) at $x = 3$

SOLUTION 2.5

a) $f(-2) = 1$ — The function at -2 is 1.

$\lim_{x \to -2} f(x)$ does not exist — The limit does not exist.

$\lim_{x \to -2} f(x) \neq 1$ — This function is *not* continuous at $x = -2$.

b) $f(-1) = 1$ — The function equals 1 at $x = -1$.

$\lim_{x \to -1} f(x) = 1$ — The limit is 1 at $x = -1$.

Therefore, the function is continuous at $x = -1$.

c) $f(0)$ = does not exist — There is *no* value for this function at $x = 0$.

$$\lim_{x \to 0} f(x) = 2$$ The limit at $x = 0$ is 2.

$f(0) \neq 2$ — This function is *not* continuous at $x = 0$.

d) $f(1) = 1$ — The function equals 1 at $x = 1$.

$$\lim_{x \to 1} f(x) = 1$$ The limit equals 1 at $x = 1$.

$1 = 1$ — The function is continuous at $x = 1$.

e) $f(3)$ = no value — The function does not exist at $x = 3$.

$\lim_{x \to 3} f(x)$ does not exist — The limit does not exist at $x = 3$.

The function value at 3 and the limit as x approaches 3 do not exist. The function is not continuous at $x = 3$.

Practice Exercises

1. Use the ideas presented in Table 2.1, and the graph for Example 2.1 to determine the values for the following.

   a) $\lim_{x \to -3} f(x)$

   b) $f(-3)$

   c) $\lim_{x \to -2} f(x)$

   d) $f(-2)$

   e) $\lim_{x \to 1} f(x)$

   f) $f(1)$

   g) $\lim_{x \to 2} f(x)$

   h) $f(2)$

   i) $\lim_{x \to 0} f(x)$

   j) $f(0)$

2. Find the limit for the following functions.

   a) $\lim_{x \to -1} (x^2 + x)$

   b) $\lim_{x \to 0} \frac{x^3 - x}{x}$

   c) $\lim_{x \to 0} \frac{\sqrt{x+2} - \sqrt{2}}{x}$

   d) $\lim_{x \to -3} \frac{x^2 + 5x + 6}{x + 3}$

   e) $\lim_{x \to 0} \frac{\frac{1}{x-2} - \frac{1}{x}}{\frac{1}{x}}$

3. Determine the limit of the following functions, if possible.

   a) $\lim_{x \to 5} \frac{1}{x - 5}$

   b) $\lim_{x \to 0} \frac{1}{x^2}$

   c) $\lim_{x \to -7} \frac{4}{(x+7)^2}$

4. Find the following limits at infinity.

   a) $\lim_{x \to \infty} \frac{x^2 - 6x + 2}{x - 5}$

   b) $\lim_{x \to \infty} \frac{3x - 6}{7x + 4}$

   c) $\lim_{x \to \infty} \frac{x + 6}{x^2 - 7}$

5. Use the graph from Example 2.5 to determine whether the function is continuous at:

   a) −2

   b) −1

   c) 0

   d) 1

   e) 3

Answers

1. a) 1/2

   b) 1/2

   c) 1

   d) does not exist

   e) does not exist

   f) does not exist

   g) 2

   h) 2

   i) 2

   j) 1

2. a) 0

   b) –1

   c) $\sqrt{2}/4$

   d) –1

   e) –1

3. a) limit does not exist

   b) ∞

   c) ∞

4. a) ∞

   b) 3/7

   c) 0

5. a) not continuous

   b) yes

   c) not continuous

   d) yes

   e) not continuous

3

The Derivative

3.1 INTRODUCTION

The derivative has many uses. There are also many techniques used to determine the derivative of a function. This chapter will cover the uses and processes involving derivatives of functions.

The topics covered will be:

a) Average rate of change
b) The concept of derivative
c) The Power Rule
d) The Chain Rule
e) The Product and Quotient rules
f) Logarithmic and exponential derivatives
g) Implicit derivatives

3.2 AVERAGE RATE OF CHANGE

A rate of change is the difference in the value of some variable during a set time interval. A common example of a rate of change is the velocity of a moving object. Velocity is the rate of change of distance with respect to time.

3.1 $$\text{Average Velocity} = \frac{\text{Distance Covered}}{\text{Elapsed Time Interval}}$$

A car that travels 120 miles from 2 p.m. to 4 p.m. has an average velocity of

$$\frac{120 \text{ miles}}{4 \text{ p.m.} - 2 \text{ p.m.}} = \frac{120 \text{ miles}}{2 \text{ hrs.}} = 60 \text{ miles per hour}$$

Table 3.1 lists the distance (in miles) two cars have traveled after leaving location *A*. The time at which these distances were reached is listed in the Time column.

	Distance from Location A	Time
car 1	10 miles 130 miles	9 a.m. 11 a.m.
car 2	30 miles 100 miles	10 a.m. 11 a.m.

Table 3.1 Time and Distance Table for Two Cars

EXAMPLE 3.1

Find the average velocity of car 1 and car 2 using the data provided in Table 3.1.

SOLUTION 3.1

a) $\text{average velocity} = \dfrac{\text{distance covered}}{\text{elapsed time}}$ Use Formula 3.1.

net distance covered = 130 – 10 miles
elapsed time = 11 a.m. – 9 a.m.
Find the net distance covered and elapsed time.

$\text{average velocity} = \dfrac{(130 - 10) \text{ miles}}{11 \text{ a.m.} - 9 \text{ a.m.}}$ Complete the substitution into Formula 3.1.

$= \dfrac{120 \text{ miles}}{2 \text{ hours}}$

$= 60$ miles per hour Complete the subtraction and division.

b) $\text{average velocity} = \dfrac{\text{distance covered}}{\text{elapsed time}}$ Use Formula 3.1.

net distance covered = 100 – 30 miles
elapsed time = 10 a.m. – 9 a.m.
Find the net distance covered and elapsed time.

$$\text{average velocity} = \frac{100 - 30 \text{ miles}}{10 \text{ a.m.} - 9 \text{ a.m.}}$$ Complete the substitution into Formula 3.1.

$$= \frac{70 \text{ miles}}{1 \text{ hour}}$$

$$= 70 \text{ miles per hour}$$ Find the answer by subtraction and division.

The *difference* in the miles traveled *divided by* the *difference* in time may remind you of two other similar formulas; the *slope* formula,

$$m = \frac{y_2 - y_1}{x_2 - x_1},$$

and the *difference quotient*,

$$\frac{f(x+h) - f(x)}{x+h-x}.$$

This similarity is not coincidental. Average rates are slopes for a measurable time interval.

3.3 THE CONCEPT OF DERIVATIVE

The technical definition of the first derivative for a function, $f(x)$ is:

> **3.2 Slope as Derivative**
>
> $$m_{\tan} = \lim_{h \to 0} \frac{f(x+h) - f(x)}{h}$$

where $m_{\tan}$ is the symbol referring to slope of a tangent line. A tangent line is a line that touches, but does not cross, a curve within a small interval on the curve.

The denominator of Formula 3.2 tends toward zero as the value of h approaches zero. If h were to reflect a time interval, it would become more instantaneous as h becomes smaller. Thus, a derivative is an instantaneous rate of change.

EXAMPLE 3.2

For the following functions use Formula 3.2 (the slope of the tangent) to find the derivative.

a) $f(x) = 5x$

b) $f(x) = \sqrt{x}$

SOLUTION 3.2

a) $f(x) = 5x$ — Copy the function.

$f(x+h) = 5(x+h)$ — Find $f(x+h)$.

$$m_{\tan} = \lim_{h \to 0} \frac{5(x+h) - 5(x)}{h}$$

Substitute the expression in $f(x)$ and $f(x+h)$ into Formula 3.2.

$$= \lim_{h \to 0} \frac{5x + 5h - 5x}{h}$$

Distribute 5 and combine like terms in numerator.

$$= \lim_{h \to 0} \frac{5h}{h}$$

$$= \lim_{h \to 0} \frac{5\cancel{h}}{\cancel{h}}$$

Cancel the h factors.

$$= \lim_{h \to 0} 5 = 5$$

The result is 5 when h approaches 0.

$$m_{\tan} = \lim_{h \to 0} \frac{5(x+h) - 5(x)}{h} = 5$$

The first derivative of $5x$ is 5.

b) $f(x) = \sqrt{x}$ — Copy the function.

$f(x+h) = \sqrt{x+h}$ — Find the function of $x + h$.

$$m_{\tan} = \lim_{h \to 0} \frac{\sqrt{x+h} - \sqrt{x}}{h}$$

Substitute the expressions for $f(x)$ and $f(x+h)$ into Formula 3.2.

$$= \lim_{h \to 0} \left(\frac{\sqrt{x+h} - \sqrt{x}}{h}\right)\left(\frac{\sqrt{x+h} + \sqrt{x}}{\sqrt{x+h} + \sqrt{x}}\right)$$

Rationalize the numerator.

$$= \lim_{h \to 0} \frac{(\sqrt{x+h})^2 - (\sqrt{x})^2}{h(\sqrt{x+h} + \sqrt{x})}$$

Complete the algebraic simplification.

$$= \lim_{h \to 0} \frac{x + h - x}{h(\sqrt{x+h} + \sqrt{x})}$$

$$= \lim_{h \to 0} \frac{h}{h(\sqrt{x+h} + \sqrt{x})}$$

$$= \lim_{h \to 0} \frac{\cancel{h}}{\cancel{h}(\sqrt{x+h}+\sqrt{x})}$$

Cancel the common factors of h.

$$= \lim_{h \to 0} \frac{1}{\sqrt{x+h}+\sqrt{x}}$$

$$= \lim_{h \to 0} \frac{1}{\sqrt{x+h}+\sqrt{x}} = \frac{1}{\sqrt{x+0}+\sqrt{x}}$$

Find the limit and add the two radical expressions in the denominator.

$$m_{\tan} = \frac{1}{\sqrt{x}+\sqrt{x}} = \frac{1}{2\sqrt{x}}$$

The derivative is $\frac{1}{2\sqrt{x}}$.

3.4 THE POWER RULE

As you may have seen from Example 3.2 the process of using the difference quotient to find a derivative can become tedious. However, there are faster ways to differentiate functions. We will cover these ways (or rules) in this chapter, beginning with the Power Rule (Formula 3.3).

> **3.3 The Power Rule**
>
> $$\text{for } f(x) = x^n$$
>
> $$f'(x) \text{ (the derivative)} = nx^{n-1}$$

The notation of $f'(x)$ is first introduced here. It means "the first derivative of $f(x)$." There are other symbols that mean "the first derivative." Table 3.2 gives these symbols.

Function Notation	Derivative Notation	Name
$f(x)$	$f'(x)$	function notation
y	y'	y-prime
y	$\frac{dy}{dx}$	Leibniz notation
an expression	D_x	operator notation

Table 3.2 Derivative Symbols

These symbols will be explained as they are introduced.

EXAMPLE 3.3

Find the derivative of the following functions.

a) $f(x) = 5$

b) $f(x) = 3x$

c) $f(x) = 7x^2 - 3x + 5$

d) $f(x) = \sqrt{x}$

e) $f(x) = 5\sqrt{x} + \dfrac{7}{\sqrt[3]{x}}$

f) $f(x) = \sqrt[3]{x^5} + 8x^7$

g) $f(x) = 30x^6$

h) $f(x) = 3x^{\sqrt{2}}$

i) $f(x) = \pi$

j) $f(x) = ax^b$ where a and b are real numbers.

SOLUTION 3.3

a) $f(x) = 5$ — Copy the function.

$f(x) = 5 \cdot 1$

$f(x) = 5x^0$

In order to more fully understand the Power Rule use $x^0 = 1$.

$f'(x) = 0 \cdot 5x^{0-1}$

$= 0 \cdot (5x^{-1})$

$= 0$

Use Formula 3.3 (Power Rule) to multiply the current exponent times the coefficient. The exponent is reduced by one.

$f(x) = 5$

$f'(x) = 0$

The first derivative of a *constant* is *zero*.

b) $f(x) = 3x$ — Copy the function.

$f(x) = 3x^1$ — Express the understood exponent of 1.

$f'(x) = 1 \cdot 3x^{1-1}$

$= 3x^0 = 3 \cdot 1 = 3$

Use Formula 3.3 to multiply the current exponent times the coefficient.

$f(x) = 3x$
$f'(x) = 3$

The first derivative is the constant, 3.

c) $f(x) = 7x^2 - 3x + 5$ — Copy the function.

$= 7x^2 - 3x^1 + 5x^0$ — Express the exponents for clarity.

$f'(x) = 2 \cdot 7x^{2-1} - 1 \cdot 3x^{1-1} + 0 \cdot 5x^{0-1}$

Use the process of the Power Rule.

$f'(x) = 14x - 3x^0 + 0$

$f'(x) = 14x - 3$ — Simplify each term.

d) $f(x) = \sqrt{x}$ — Copy the function.

$f(x) = x^{1/2}$ — Rewrite the radical in fractional exponent form.

$f'(x) = \frac{1}{2}x^{1/2 - 1}$ — Use the process of the Power Rule.

$f'(x) = \frac{1}{2}x^{1/2 - 2/2}$

$f'(x) = \frac{1}{2}x^{-1/2}$

$f'(x) = \frac{1}{2} \cdot \frac{1}{x^{1/2}} = \frac{1}{2} \cdot \frac{1}{\sqrt{x}}$ — Rewrite the derivative in radical form.

$= \frac{1}{2\sqrt{x}}$

e) $f(x) = 5\sqrt{x} - \dfrac{7}{\sqrt[3]{x}}$	Copy the function.
$f(x) = 5x^{1/2} - \dfrac{7}{x^{1/3}}$	Rewrite the radicals as fractional exponents.
$f(x) = 5x^{1/2} - 7x^{-1/3}$	Convert the last term from a fraction to a negative exponent.
$f'(x) = \dfrac{1}{2}(5)x^{\frac{1}{2}-1} - \left(-\dfrac{1}{3}\right)7x^{-\frac{1}{3}-1}$	Use the Power Rule.
$f'(x) = \dfrac{5}{2}x^{-\frac{1}{2}} + \dfrac{7}{3}x^{-\frac{4}{3}}$	Simplify the terms.
$f'(x) = \dfrac{5}{2}\dfrac{1}{x^{1/2}} + \dfrac{7}{3}\dfrac{1}{x^{4/3}}$	Rewrite negative exponents as reciprocals.
$f'(x) = \dfrac{5}{2\sqrt{x}} + \dfrac{7}{3\sqrt[3]{x^4}}$	Convert fractional exponents into radical form.
f) $f(x) = \sqrt[3]{x^5} + 8x^7$	Copy the function.
$f(x) = x^{5/3} + 8x^7$	Rewrite the radical in fractional exponent form.
$f'(x) = \dfrac{5}{3}x^{\frac{5}{3}-1} + 7(8)x^{7-1}$	Use the Power Rule.
$f'(x) = \dfrac{5}{3}x^{2/3} + 56x^6$	Simplify the terms.
$f'(x) = \dfrac{5}{3}\sqrt[3]{x^2} + 56x^6$	Convert the fractional exponent to radical form.
g) $f(x) = 30x^6$	Copy the function.
$f'(x) = 6(30)x^{6-1}$	Use the Power Rule.
$f'(x) = 180x^5$	Simplify the function.
h) $f(x) = 3x^{\sqrt{2}}$	Copy the function.
$f'(x) = \sqrt{2}(3)(x^{\sqrt{2}-1})$	Use the Power Rule.

$f'(x) = 3\sqrt{2}x^{\sqrt{2}-1}$ — Simplify the function as much as possible.

i) $f(x) = \pi$ — Copy the function.

$f'(x) = 0$
because π is a constant. — Use the Power Rule. Even though π is an unusual symbol, it stands for a constant roughly equal to $22/7$. The derivative is 0.

j) $f(x) = ax^b$ — Copy the function.

$f'(x) = (b)(a)x^{b-1}$ — Use the Power Rule. (Remember *a* and *b* are numbers.)

$f'(x) = abx^{b-1}$

3.5 THE GENERAL POWER RULE (CHAIN RULE FOR POWERS)

As functions get more complex in nature, it may be useful to use a more versatile rule. The General Power Rule is useful for composite (nested) functions.

3.4 The General Power Rule
Given: $f(x) = [g(x)]^n$
Then: $f'(x) = n[g(x)]^{n-1}(g'(x))$

The most important aspect of using this rule is to concentrate on the exponent *first*, then multiply by the derivative of the function that the exponent governed.

For example, the following function has a binomial $(x^2 + 3x)$ raised to the 4th power.

$$f(x) = (x^2 + 3x)^4$$

Finding the derivative of this function begins by treating the function as if it were any variable to the 4th power. After the exponent has been moved according to the Power Rule, then the binomial inside the parentheses can be differentiated. Figure 3.1 shows the thought process involved in the differentiation of the entire function.

$f(x) = (x^2 + 3x)^4$	
$f'(x) = 4(\quad)^3$	Begin with the Power Rule. Do *not* focus on contents of parentheses.
$f'(x) = 4(x^2 + 3x)^3$	*Now* turn your attention to the binomial.
$f'(x) = 4(x^2 + 3x)^3 \cdot (2x + 3)$ — derivative of $x^2 + 3x$	Multiply by the derivative of the binomial.
$f'(x) = 4(2x + 3)(x^2 + 3x)^3$	Rearranging the factors completes the derivative.

Figure 3.1 The General Power Rule

EXAMPLE 3.4

Find the derivative of the following functions. Use the General Power Rule.

a) $f(x) = (3x^2 + 7x + 1)^5$

b) $f(x) = 4(5 - x)^6$

c) $f(x) = \sqrt[3]{x + 1}$

d) $f(x) = \dfrac{5}{x^2 + 2x + 4}$

e) $f(x) = \dfrac{3}{\sqrt{x^2 + 2x}}$

SOLUTION 3.4

a) $f(x) = (3x^2 + 7x + 1)^5$ — Copy the function.

$f(x) = (3x^2 + 7x + 1)^{\mathbf{5}} = [g(x)]^{\mathbf{5}}$ — Note that this function is an expression to the 5th power.

$f'(x) = \mathbf{5}(3x^2 + 7x + 1)^{\mathbf{4}}(\mathbf{6x + 7})$ — Use the General Power Rule. The bold letters and numbers are changes through differentiation.

$\qquad\qquad g(x) \qquad\qquad g'(x)$

$f'(x) = 5(6x + 7)(3x^2 + 7x + 1)^4$ — Move the binomial for clarity.

b) $f(x) = 4(5 - x)^6$ — Copy the function.

$f(x) = 4(5 - x)^{\mathbf{6}} = 4[g(x)]^{\mathbf{6}}$ — Note that this function is a constant times an expression to the 6th power.

$f'(x) = \mathbf{6} \cdot 4(5 - x)^{\mathbf{5}}(\mathbf{-1})$ — Use the General Power Rule. The bold typeface shows changes.

$\qquad\qquad g(x) \qquad g'(x)$

$f'(x) = -24(5 - x)^5$ — Simplify.

c) $f(x) = \sqrt[3]{x + 1}$ — Copy the function.

$f(x) = (x + 1)^{1/3}$ — Rewrite the radical as a fractional exponent.

$f(x) = (x + 1)^{\mathbf{1/3}} = [g(x)]^{\mathbf{1/3}}$ — Note that this function is an expression to the 1/3 power.

$f'(x) = \frac{1}{3}(x + 1)^{\frac{1}{3} - 1}(1)$ — Use the General Power Rule. The bold typeface shows changes.

$\quad = \mathbf{\frac{1}{3}}(x + 1)^{\mathbf{-2/3}}(\mathbf{1})$

$\qquad\qquad g(x) \qquad g'(x)$

$f'(x) = \frac{1}{3}(x+1)^{-\frac{2}{3}}$ Simplify the function.

$f'(x) = \frac{1}{3}\frac{1}{\sqrt[3]{(x+1)^2}} = \frac{1}{3\sqrt[3]{(x+1)^2}}$ Rewrite as a radical expression.

d) $f(x) = \frac{5}{x^2+2x+4}$ Copy the function.

$f(x) = 5(x^2+2x+4)^{-1}$ Rewrite the fraction as a negative exponent.

$f(x) = 5(x^2+2x+4)^{\mathbf{-1}} = 5[g(x)]^{\mathbf{-1}}$

Note that the function is a constant times an expression to the –1 power.

$f'(x) = \mathbf{(-1)}\ (5)\ (x^2+2x+4)^{-1\mathbf{-1}}\ \mathbf{(2x+2)}$

Use the General Power Rule. The bold typeface shows the changes.

$f'(x) = -5(2x+2)(x^2+2x+4)^{-2}$ Simplify the function.

$f'(x) = \frac{-5(2x+2)}{(x^2+2x+4)^2}$ Rewrite the expression as a rational function.

$f'(x) = \frac{-10(x+1)}{(x^2+2x+4)^2}$ Factor the numerator.

e) $f(x) = \frac{3}{\sqrt{x^2+2x}}$ Copy the function.

$f(x) = \frac{3}{(x^2+2x)^{1/2}}$ Rewrite the radical expression as an expression with a rational exponent.

$f(x) = 3(x^2+2x)^{-\frac{1}{2}}$ Rewrite the fraction using a negative exponent.

$f(x) = 3[g(x)]^{\mathbf{-\frac{1}{2}}}$ Note the function is a constant times an expression to the $-1/2$ power.

$$f'(x) = \left(-\frac{1}{2}\right)(3)(x^2+2x)^{-\frac{1}{2}-\mathbf{1}}\ (\mathbf{2x+2})$$

Use the General Power Rule. The bold typeface shows the changes.

$$f'(x) = -\frac{3}{2}(x^2+2x)^{-\frac{3}{2}}(2x+2)$$

$$f'(x) = -\frac{3}{2}(2x+2)(x^2+2x)^{-\frac{3}{2}}$$

Simplify the expression.

$$f'(x) = (-3x-3)(x^2+2x)^{-\frac{3}{2}}$$

$$f'(x) = \frac{-3x-3}{\sqrt{(x^2+2x)^3}}$$

Rewrite negative exponent as radical expression.

$$f'(x) = \frac{-3(x+1)}{\sqrt{(x^2+2x)^3}}$$

Factor the expression.

3.6 THE PRODUCT AND QUOTIENT RULES

The derivative of a product of two functions is, unfortunately, *not* the product of their respective derivatives. Similarly, the quotient of the derivative is *not* the derivative of the quotient.

The rules for products and quotients are described in the following text.

The Product Rule

3.5 The Product Rule

Given: $f(x) = F(x) \cdot S(x)$

Then $f'(x) = F'(x)S(x) + F(x)S'(x)$

$F(x)$ = first factor, $S(x)$ = second factor

The Product Rule can be stated in a less symbolic form. If the original function, $f(x)$, can be described as a product of two factors, then the derivative of the function, $f'(x)$, is

(first factor derivative)(second factor) + (first factor)(second factor derivative)

The function, $f(x) = (x^2 - 1)(3x^3 + 2x)$, can be differentiated using the product rule. It is often helpful to list each factor with its respective derivative immediately below it.

	First	Second
Factor:	$x^2 - 1$ $F(x)$	$3x^3 + 2x$ $S(x)$
Derivative of Factor:	$2x$ $F'(x)$	$9x^2 + 2$ $S'(x)$

Pair each derivative with the other factor to get the complete derivative of $f(x)$.

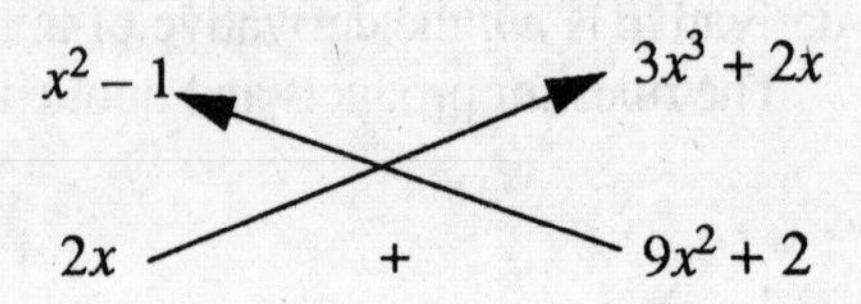

Therefore: $f'(x) = \underbrace{[2x]}_{F'(x)}\ \underbrace{(3x^3 + 2x)}_{S(x)} + \underbrace{(x^2 - 1)}_{F(x)}\ \underbrace{[9x^2 + 2]}_{S'(x)}$

The Quotient Rule

3.6 The Quotient Rule

$$\text{For } f(x) = \frac{T(x)}{B(x)}$$

$$f'(x) = \frac{T'(x)B(x) - T(x)B'(x)}{(B(x))^2}$$

where $T(x)$ is the numerator (top) function and $B(x)$ is the denominator (bottom) function.

The less symbolic statement for the quotient rule of $f(x)$ is

$$f'(x) = \frac{(\text{derivative of top})(\text{bottom}) - (\text{top})(\text{derivative of bottom})}{(\text{bottom})^{\text{squared}}}$$

The function, $f(x) = \dfrac{3x^2+7}{x^3+5x}$ can be differentiated using the quotient rule. As with the Product Rule, the numerator and the denominator can be listed with their respective derivatives.

	Numerator	**Denominator**
Function	$3x^2+7$ $T(x)$	x^3+5x $B(x)$
Derivative	$6x$ $T'(x)$	$3x^2+5$ $B'(x)$

The product pairs are $(6x)(x^3+5x)$ and $(3x^2+7)(3x^2+5)$. The numerator of the **derivative** is $(6x)(x^3+5x) - (3x^2+7)(3x^2+5)$. It is formed by subtracting the product pairs in the *proper* order.

The total derivative, which is itself a function, is

$$f'(x) = \frac{(6x)(x^3+5x) - (3x^2+7)(3x^2+5)}{(x^3+5x)^2}$$

After simplifying the numerator, the derivative becomes:

$$f'(x) = \frac{-(3x^4 + 6x^2 + 35)}{(x^3 + 5x)^2}$$

One of the most frustrating aspects of the Product Rule and the Quotient Rule is not the actual differentiation, but the factoring and simplification needed after the differentiation is finished. In the next example, the simplification process will be given in detail.

EXAMPLE 3.5

Find the derivatives of the following functions. Use the General Power Rule where necessary.

a) $f(x) = (x^2 + 1)(3x - 1)$

b) $f(x) = \dfrac{(2x-3)}{(5x+1)}$

c) $f(x) = (3x+1)^5 (x^2 - x)^4$

d) $f(x) = \dfrac{(2x-5)^4}{(4x-2)^3}$

e) $f(x) = (2x-1)^{1/2} (3x+5)^{1/3}$

f) $f(x) = \sqrt{(6x+1)(x^2 - 5)}$

g) $f(x) = \sqrt[3]{\dfrac{(9x+2)}{(3x^2+5)}}$

h) $f(x) = \dfrac{3}{\sqrt{x}}$

SOLUTION 3.5

a) $f(x) = (x^2 + 1)(3x - 1)$ — Copy the function.

Factor	$(x^2 + 1)$	$(3x - 1)$	List each factor and its
Derivative	$(2x)$	(3)	derivative.

$$f'(x) = \mathbf{(2x)}\ (3x-1) + (x^2+1)\ \mathbf{(3)}$$

Multiply each factor by the derivative of the other factor. Each derivative is in boldface type.

$$f'(x) = 6x^2 - 2x + 3x^2 + 3$$

Simplify by distributing factors.

$$f'(x) = 9x^2 - 2x + 3$$

Combine like terms.

b) $f(x) = \dfrac{(2x-3)}{(5x+1)}$

Copy the function.

	numerator	denominator
Factor	$(2x-3)$	$(5x+1)$
Derivative	(2)	(5)

List each factor and the derivative of each.

$$f'(x) = \frac{\mathbf{(2)}\ (5x+1) - \mathbf{(5)}\ (2x-3)}{(5x+1)^2}$$

Multiply each factor by the derivative of the other (bold type). Divide by the square of the denominator.

$$f'(x) = \frac{10x + 2 - 10x + 15}{(5x+1)^2}$$

Distribute the factors of the derivative numerator. (Observe the change of sign in the last two terms.)

$$f'(x) = \frac{17}{(5x+1)^2}$$

Combine like terms for the final answer. It is not necessary to multiply and expand the denominator.

c) $f(x) = (3x+1)^5 (x^2-x)^4$

Copy the function.

Factor	$(3x+1)^5$	$(x^2-x)^4$
Derivative	$5(3x+1)^4(3)$	$4(x^2-x)^3(2x-1)$

List each factor and its derivative.

$$f'(x) = [\mathbf{5\,(3x+1)^4\,(3)}]\,(x^2-x)^4 + \left[\mathbf{4\,(x^2-x)^3\,(2x-1)}\right](3x+1)^5$$

Multiply each factor by the derivative of the other factor. (Each derivative is in boldface type.)

$$f'(x) = 15\,(3x+1)^4\,(x^2-x)^4 + (8x-4)\,(x^2-x)^3\,(3x+1)^5$$

Simplify the terms.
$3 \cdot 5 = 15,$
$4(2x-1) = 8x-4.$

$$f'(x) = (3x+1)^4\,(x^2-x)^3\,[15\,(x^2-x) + (8x-4)\,(3x+1)]$$

Factor out 4 factors of $(3x+1)$ and 3 factors (x^2-x).

$$f'(x) = (3x+1)^4\,(x^2-x)^3\,[15x^2 - 15x + 24x^2 - 4x - 4]$$

Multiply.

$$f'(x) = (3x+1)^4\,(x^2-x)^3\,(39x^2 - 19x - 4)$$

Combine like terms for final answer.

d) $f(x) = \dfrac{(2x-5)^4}{(4x-2)^3}$

Copy the function.

	numerator	denominator
Factor	$(2x-5)^4$	$(4x-2)^3$
Derivative	$4\,(2x-5)^3\,(2)$	$3\,(4x-2)^2\,(4)$

List the numerator and denominator and the derivative of each.

$$f'(x) = \frac{\mathbf{4\,(2x-5)^3\,(2)}\,(4x-2)^3 - (2x-5)^4\,\mathbf{(3)\,(4x-2)^2\,(4)}}{[(4x-2)^3]^2}$$

Multiply each factor by the derivative of the other factor (bold type). Divide by the denominator squared.

$$f'(x) = \frac{8(2x-5)^3(4x-2)^1 - 12(2x-5)^4}{(4x-2)^4}$$

Cancel two factors of $(4x-2)$ from each term and the denominator.

$$f'(x) = \frac{4(2x-5)^3[2(4x-2) - 3(2x-5)]}{(4x-2)^4}$$

Factor out 3 factors of $(2x-5)$ and a factor of 4.

$$f'(x) = \frac{4(2x-5)^3(8x-4-6x+15)}{(4x-2)^4}$$

Multiply. Watch minus signs!

$$f'(x) = \frac{4(2x-5)^3(2x+11)}{(4x-2)^4}$$

Combine like terms for the final answer.

e) $f(x) = (2x-1)^{1/2}(3x+5)^{1/3}$

Copy the function.

	numerator	denominator
Factor	$(2x-1)^{1/2}$	$(3x+5)^{1/3}$
Derivative	$\frac{1}{2}(2x-1)^{-1/2}(2)$	$\frac{1}{3}(3x+5)^{-2/3}(3)$

List the numerator and denominator and the derivative of each.

$$f'(x) = \left[\frac{1}{2}(2x-1)^{-1/2}(2)\right](3x+5)^{1/3} + \left[\frac{1}{3}(3x+5)^{-2/3}(3)\right](2x-1)^{1/2}$$

Multiply each factor by the derivative of the other factor.

$$f'(x) = (2x-1)^{-1/2}(3x+5)^{1/3} + (3x+5)^{-2/3}(2x-1)^{1/2}$$

Multiply the constants.

$$f'(x) = (2x-1)^{-1/2}(3x+5)^{-2/3}[(3x+5)^1 + (2x-1)^1]$$

Factor out the binomials with the negative exponents. This causes the power of the other factors to become 1.

$$f'(x) = (2x-1)^{-1/2}(3x+5)^{-2/3}(5x+4)$$

Combine similar terms.

$$f'(x) = \frac{5x+4}{(2x-1)^{1/2}(3x+5)^{2/3}}$$

This gives the final answer. The negative exponents have been rewritten in the denominator.

f) $f(x) = \sqrt{(6x+1)(x^2-5)}$

Copy the function.

$$f(x) = [(6x+1)(x^2-5)]^{1/2}$$

Rewrite the radical as a fractional exponent.

$$f(x) = (6x+1)^{1/2}(x^2-5)^{1/2}$$

Distribute exponents among separate factors.

	numerator	denominator
Factor	$(6x+1)^{1/2}$	$(x^2-5)^{1/2}$
Derivative	$\frac{1}{2}(6x+1)^{-1/2}(6)$	$\frac{1}{2}(x^2-5)^{-1/2}(2x)$

List each factor and its derivative.

$$f'(x) = \left[\mathbf{\frac{1}{2}(6x+1)^{-1/2}(6)}\right](x^2-5)^{1/2} + (6x+1)^{1/2}\left[\mathbf{\frac{1}{2}(x^2-5)^{-1/2}(2x)}\right]$$

Multiply each factor by the derivative of the other (bold type).

$$f'(x) = 3(6x+1)^{-1/2}(x^2-5)^{1/2} + x(6x+1)^{1/2}(x^2-5)^{-1/2}$$

Multiply constants and first degree factors. $\frac{1}{2} \cdot 6 = 3$, $\frac{1}{2} \cdot 2x = x$.

$$f'(x) = (6x+1)^{-1/2}(x^2-5)^{-1/2}[3(x^2-5) + x(6x+1)]$$

Factor out binomials with negative exponents. The powers of the remaining factors rise to 1.

$$f'(x) = (6x-1)^{-1/2}(x^2-5)^{-1/2}(3x^2-15+6x^2+x)$$

Multiply and distribute.

$$f'(x) = (6x-1)^{-1/2}(x^2-5)^{-1/2}(9x^2+x-15)$$

Combine like terms.

$$f'(x) = \frac{9x^2+x-15}{(6x-1)^{1/2}(x^2-5)^{1/2}}$$

Place factors with negative exponents into denominator.

g) $$f(x) = \sqrt[3]{\frac{(9x+2)}{(3x^2+5)}}$$

Copy the function.

$$f(x) = \frac{\sqrt[3]{9x+2}}{\sqrt[3]{3x^2+5}}$$

Use the rules of radical expressions to separate numerator and denominator.

$$f(x) = \frac{(9x+2)^{1/3}}{(3x^2+5)^{1/3}}$$

Rewrite the radical expressions with fractional exponents.

	numerator	denominator
Factor	$(9x+2)^{1/3}$	$(3x^2+5)^{1/3}$
Derivative	$\frac{1}{3}(9x+2)^{-2/3}(9)$	$\frac{1}{3}(3x^2+5)^{-2/3}(6x)$

List each factor and its derivative.

$$f'(x) = \frac{\left[\mathbf{\frac{1}{3}(9x+2)^{-2/3}(9)}\right](3x^2+5)^{1/3} - \left[\mathbf{\frac{1}{3}(3x^2+5)^{-2/3}(6x)}\right](9x+2)^{1/3}}{((3x^2+5)^{1/3})^2}$$

Multiply the numerator and denominator by the other's derivative (bold type). Place these over the denominator squared.

$$f'(x) = \frac{3(9x+2)^{-2/3}(3x^2+5)^{1/3} - 2x(9x+2)^{1/3}(3x^2+5)^{-2/3}}{(3x^2+5)^{2/3}}$$

Multiply constants and first degree terms. $\frac{1}{3} \cdot 9 = 3$, $\frac{1}{3} \cdot 6x = 2x$. (The $1/3$ power squared is $2/3$.)

$$f'(x) = \frac{(9x+2)^{-2/3}(3x^2+5)^{2/3}[3(3x^2+5) - 2x(9x+2)]}{(3x^2+5)^{2/3}}$$

Remove the binomials with negative exponents. The remaining binomials have powers of 1.

$$f'(x) = \frac{(9x+2)^{-2/3}(3x^2+5)^{-2/3}[9x^2+15-18x^2-4x]}{(3x^2+5)^{2/3}}$$

Distribute factors. Watch minus sign!

$$f'(x) = \frac{(9x+2)^{-2/3}(3x^2+5)^{-2/3}[-9x^2+15-4x]}{(3x^2+5)^{2/3}}$$

Combine like terms.

$$f'(x) = \frac{-9x^2-4x+15}{(9x+2)^{2/3}(3x^2+5)^{2/3}(3x^2+5)^{2/3}}$$

Place the factors with the negative exponents in the denominator.

$$f'(x) = \frac{-9x^2-4x+15}{(9x+2)^{2/3}(3x^2+5)^{4/3}}$$

Add exponents of common base $(3x^2+5)$ to obtain final result.

h) $f(x) = \dfrac{3}{\sqrt{x}}$

Copy the function.

$$f(x) = 3x^{-1/2}$$

$$f'(x) = -\frac{1}{2}(3)x^{-1/2-1}$$

$$f'(x) = -\frac{3}{2}x^{-3/2}$$

In the previous section, we found that we could rewrite the single terms in the denominator as terms with negative exponents.

Radicals can be written as fractional exponents. With this done, we use the Power Rule to obtain the derivative. Let's rework the problem using the Quotient Rule.

	numerator	denominator
Factor	3	$x^{1/2}$
Derivative	0	$\frac{1}{2}x^{-1/2}$

List the numerator and denominator and their derivatives.

$$f'(x) = \frac{(\mathbf{0}) \cdot x^{1/2} - 3\left(\frac{1}{2}\right)(\mathbf{x^{-1/2}})}{(x^{1/2})^2}$$

Multiply each fraction by the derivative of the other (bold type). Place over the denominator squared.

$$f'(x) = \frac{0 - \frac{3}{2}x^{-1/2}}{x^1}$$

Simplify the numerator and denominator.

$$f'(x) = -\frac{3}{2}x^{-1/2} \cdot x^{-1}$$

When the denominator is moved to the numerator, the exponent becomes negative.

$$f'(x) = -\frac{3}{2}x^{-1/2 - 1}$$

$$f'(x) = -\frac{3}{2}x^{-3/2}$$

Multiply the expressions with the same base by adding exponents. The result is the *same* as that from using the Power Rule.

Note: The last example shows that these rules are interrelated. If you should fail to see that the derivative of example (f) can be done by the Power Rule, you can get the same result by using the Quotient Rule. The latter method may be more lengthy, but it is still valid.

3.7 THE DERIVATIVE OF ln$x$ AND e^x

Both the natural logarithm (ln) and the natural number (e) have business and life science applications. They both also have unique derivatives. The basic derivatives for the functions e^x and $\ln x$ are listed in the rules below.

3.7 The Derivative of e^x

Given: $f(x) = e^x$

Then: $f'(x) = e^x$

The derivative of e^x, then, is simply e^x.

3.8 The Derivative of $\ln x$

Given: $f(x) = \ln x$

Then: $f'(x) = \frac{1}{x}$

These rules can be extended to situations where the exponent of e^x or the argument of $\ln x$ are not the single variable, x, but a function of x. We can then find the derivatives for $f(x) = e^{[g(x)]}$ and $f(x) = \ln [g(x)]$, where $g(x)$ is any function of x. The rules are given below.

3.9 Derivative for $e^{[g(x)]}$

Given: $f(x) = e^{[g(x)]}$

Then: $f'(x) = e^{[g(x)]} \cdot [g'(x)]$

$g(x)$ is a function of x

The above rule can be loosely paraphrased by stating that the derivative of e^{function} is $(e^{\text{function}}) \cdot$ (derivative of the function).

The rule for the natural logarithm of a function follows a similar pattern.

3.10 Derivative of $\ln[g(x)]$

Given: $f(x) = \ln[g(x)]$

Then: $f'(x) = \dfrac{1}{g(x)} \cdot [g'(x)]$

Thus, the natural logarithm of a function has a derivative that is (the reciprocal of the function) times (the derivative of the function).

EXAMPLE 3.6

Find the derivative of the following functions.

a) $f(x) = 3e^{x}$

b) $f(x) = 5\ln x$

c) $f(x) = 2e^{x^2-1}$

d) $f(x) = \ln(x^3 - 2x + 2)$

e) $f(x) = e^{\frac{5x}{x-3}}$

f) $f(x) = \ln \dfrac{x-2}{2x+3}$

g) $f(x) = \ln(x-1)^5$

h) $f(x) = \ln(\ln x)$

i) $f(x) = \dfrac{e^{x}}{\ln x}$

j) $f(x) = e^{-x} + e^{x}$

k) $f(x) = 3\ln x^2 + 5e^{x^2-2}$

SOLUTION 3.6

a) $f(x) = 3e^x$	Copy the function.
$f'(x) = 3e^x(1)$ $= 3e^x$	The numerical coefficient of e^x is not affected in the derivative process if the exponent is x. The derivative of x is 1 and $1 \cdot 3 = 3$.
b) $f(x) = 5\ln x$	Copy the function.
$f'(x) = 5\left(\frac{1}{x}\right)$ $f'(x) = \frac{5}{x}$	The derivative of the natural logarithm $\left(\frac{1}{x}\right)$ is multiplied by 5.
c) $f(x) = 2e^{x^2 - 1}$	Copy the function.
$g(x) = x^2 - 1;\ g'(x) = 2x$	Here, $g(x) = x^2 - 1$. List the derivative.
$f'(x) = 2e^{x^2 - 1}(\mathbf{2x})$	Multiply the original function by the derivative of $g(x)$ (bold type).
$f'(x) = 4xe^{x^2 - 1}$	Simplify.
d) $f(x) = \ln(x^3 - 2x + 2)$	Copy the function.
$g(x) = x^3 - 2x + 2$ $g'(x) = 3x^2 - 2$	List $g(x)$ and its derivative.
$f'(x) = \frac{\mathbf{1}}{\mathbf{x^3 - 2x + 2}} \cdot (\mathbf{3x^2 - 2})$	Multiply the reciprocal of $g(x)$ by the derivative of $g(x)$.
$f'(x) = \frac{3x^2 - 2}{x^3 - 2x + 2}$	Simplify.

e) $f(x) = e^{\frac{5x}{x-3}}$	Copy the function.
$g(x) = \dfrac{5x}{x-3}$	List $g(x)$ and its derivative.
$g'(x) = \dfrac{5(x-3)-(5x)(1)}{(x-3)^2} = -\dfrac{15}{(x-3)^2}$	
$f'(x) = \left(e^{\frac{5x}{x-3}}\right)\left(\dfrac{\mathbf{-15}}{\mathbf{(x-3)^2}}\right)$	Multiply the original function by the derivative of $g(x)$ **(bold type)**.
$f'(x) = -\dfrac{15e^{\frac{5x}{x-3}}}{(x-3)^2}$	Simplify.
f) $f(x) = \ln\dfrac{x-2}{2x+3}$	Copy the function.
$f(x) = \ln(x-2) - \ln(2x+3)$	Rewrite the function using the rules of logarithms.
function: $(x-2)$, $(2x-3)$; derivative: (1), (2)	List both $(x-2)$ and $(2x+3)$ and the derivatives of each.
$f'(x) = \dfrac{1}{x-2}\mathbf{(1)} - \dfrac{1}{2x+3}\mathbf{(2)}$	The derivative is the reciprocal of each function times its derivative.
$f'(x) = \dfrac{1}{x-2} - \dfrac{2}{2x+3}$	
$f'(x) = \dfrac{(2x+3)(1)-2(x-2)}{(x-2)(2x+3)}$	Rewrite as a single fraction, if desired.
$f'(x) = \dfrac{7}{(x-2)(2x+3)}$	
g) $f(x) = \ln(x-1)^5$	Copy the function.
$f(x) = 5\ln(x-1)$	Use the rules of logarithms to rewrite the function.

$g(x) = x-1,\ g'(x) = 1$	List $g(x)$ and its derivative.
$f'(x) = 5\left(\dfrac{1}{x-1}\right)(1)$	Multiply the reciprocal of $g(x)$ by its derivative.
$f'(x) = \dfrac{5}{x-1}$	
h) $f(x) = \ln(\ln x)$	Copy the function.
$g(x) = \ln x$	We must use $g(x) = \ln x$ here.
$g'(x) = \dfrac{1}{x}$	List $g(x)$ and its derivative.
$f'(x) = \left(\dfrac{1}{\ln x}\right)\cdot\left(\dfrac{1}{x}\right)$	Multiply the reciprocal of $g(x)$ by its derivative.
$f'(x) = \dfrac{1}{x\ln x}$	Simplify.
i) $f(x) = \dfrac{e^x}{\ln x}$	Copy the function.
	This solution involves the Quotient Rule.

	numerator	denominator
factor	e^x	$\ln x$
derivative	e^x	$\dfrac{1}{x}$

List the numerator and denominator and their derivatives.

$f'(x) = \dfrac{e^x \ln x - e^x\left(\dfrac{1}{x}\right)}{(\ln x)^2}$	Multiply the numerator and the denominator by the other's derivative. Place this over the denominator squared.
$f'(x) = \dfrac{e^x\left(\ln x - \dfrac{1}{x}\right)}{(\ln x)^2}$	Simplify by factoring out e^x.

j) $f(x) = e^{-x} + e^{x}$	Copy the function.
$f'(x) = e^{-x}(-1) + e^{x}$	Each term follows the rules for the derivative of e^x.
$f'(x) = -e^{-x} + e^{x}$	Simplify.
k) $f(x) = 3\ln x^2 + 5e^{x^2-2}$	Copy the function.
$f(x) = 3(2)\ln x + 5e^{x^2-2}$	Use the rules of logarithms to rewrite the first term.
$f(x) = 6\ln x + 5e^{x^2-2}$	
$g(x) = x^2 - 2;\ g'(x) = 2x$	The first term is a constant times ln*x*, and the basic derivative of ln*x* applies. The second term has a function (x^2-2) for the exponent of *e*. Let $g(x) = (x^2-2)$. Then $g'(x) = 2x$.
$f'(x) = 6\left[\mathbf{\frac{1}{x}}\right] + 5e^{x^2-2}[\mathbf{2x}]$	The bold type shows changes occurring through differentiation.
$f'(x) = \frac{6}{x} + 10xe^{x^2-2}$	Simplify.

3.8 IMPLICIT DIFFERENTIATION

There are some algebraic expressions that cannot be defined as *y* in terms of *x*. Most of the examples discussed in this text have been expressed as a function of *x* — that is, $f(x)$. If we were to let $y = f(x)$, the expressions would then show *y* in terms of *x*. For $f(x) = 3x^2 - 7x + 2$ to be expressed as *y* in terms of *x*, we rewrite the function as $y = 3x^2 - 7x + 2$. Note that a single *y*-variable with a power of 1 is equal to a combination of *x*-variables.

The expression $y^3 - 2y = 6 - x$ presents a different problem. No matter what algebraic processes we use, we cannot get a single, first-power *y* to equal some combination of *x*-variables. However, this problem does *not* stop us from finding an *implied derivative*. The derivative of y is "assumed" to be found when we place y' in the equation. The symbol for the derivative, y' (y-prime), is often used to indicate that a derivative has been found.

Thus,

$$y^3 - 2y = 6 - x$$

is differentiated to result in

$$3y^2y' - 2y' = 0 - 1$$

where y = {some unknown function of x}
y' = {the derivative of the unknown function of x}.

One difficulty that occurs when working with implicit derivatives is the nature of y. The variable, y, is really a function. (This function could be large and complex — if we could determine it.) We must treat y as we would $\sqrt{x+2}$ or e^x or $\ln(x-5)$ or $(x^2-5)^6$. Therefore, the Product, Quotient, and Chain rules often apply.

Before we examine implicit differentiation through some examples, we will look at a four-step process used to complete these derivatives.

Stepwise Process for Implicit Derivatives

1. *Differentiate* — using applicable rules.
2. *Isolate* the y' terms from the terms not containing y'.
3. *Factor* out y' (if necessary).
4. *Divide* (if necessary).

EXAMPLE 3.7

Find the following derivatives through implicit differentiation.

a) $y^2 + 3y = x$
b) $y^2 = e^x$
c) $xy = 4$
d) $xy^2 - 2y = 6x$
e) $\sqrt{y} - 2x = 10$
f) $e^{xy} = 12x$

g) $\ln xy = 10x$

h) $x^2 + y^2 = 16$

SOLUTION 3.7

a) $y^2 + 3y = x$	Copy the expression.
$2yy' + 3y' = 1$	Differentiate, using the Chain Rule on y^2. The derivative of y with respect to x is y' and the derivative of x with respect to x is 1. The implied derivatives y' are in bold type.
$2yy' + 3y' = 1$	The y's are already isolated on one side of the equals sign.
$y'(2y+3) = 1$	Factor out y'.
$y' = \dfrac{1}{2y+3}$	Divide for the final answer.
b) $y^2 = e^x$	Copy the expression.
$2yy' = e^x$	Differentiate, using the Chain Rule on y^2, because y^2 is treated as a function.
$y' = \dfrac{e^x}{2y}$	Divide to complete the solution.
c) $xy = 4$	Copy the expression.
$1 \cdot y + x \cdot y' = 0$ $F'(x)S(x) + F(x)S'(x)$	Differentiate. The xy-term is a product of x times the function y. Use the Product Rule. The derivative of 4 is 0.
$xy' = -1y$	Isolate the y' term.

$$y' = \frac{-1y}{x} = \frac{-y}{x}$$

No factoring is necessary. Divide by x to obtain the final answer.

d) $xy^2 - 2y = 6x$

Copy the expression.

$$1 \cdot y^2 + x \cdot 2yy' - 2y' = 6$$

$$F'(x)S(x) + F(x)S'(x)$$

Differentiate, using the Product Rule on the xy^2 term. The derivative of $2y$ is $2y'$ and the derivative of $6x$ is 6.

$$y^2 + 2xyy' - 2y' = 6$$

Simplify.

$$2xyy' - 2y' = 6 - y^2$$

Isolate the y'-terms on one side of the equal sign. Subtract y^2.

$$y'(2xy - 2) = 6 - y^2$$

Factor out y'.

$$y' = \frac{6 - y^2}{2xy - 2}$$

Divide to complete the solution.

e) $\sqrt{y} - 2x = 10$

Copy the expression.

$$y^{1/2} - 2x = 10$$

Rewrite $\sqrt{y}$ as $y^{1/2}$.

$$\frac{1}{2}y^{-1/2}y' - 2 = 0$$

Differentiate $y^{1/2}$ as if it were a function to the $1/2$ power. The derivatives of $-2x$ and of 10 are -2 and 0, respectively.

$$\frac{1}{2}y^{-1/2}y' = 2$$

Isolate the y'-term.

$$y' = \frac{2}{\frac{1}{2}y^{-1/2}}$$

Factoring is not necessary. Divide.

$$y' = \frac{2y^{1/2}}{1/2} = \frac{2}{2} \cdot \frac{2y^{1/2}}{1/2} = 4y^{1/2}$$

Simplify by placing $y^{1/2}$ in the numerator and multiplying by the fraction $2/2$.

$y' = 4\sqrt{y}$ — Complete by converting to radical form.

f) $e^{xy} = 12x$ — Copy the expression.

$e^{xy}(1 \cdot y + xy') = 12$ — Differentiate. Use the Chain Rule for e^x. The xy-term requires the Product Rule. The derivative of $12x$ is 12.

$e^{xy}(y + xy') = 12$ — Isolate the y'-term by dividing by e^{xy} and subtracting y.

$$y + xy' = \frac{12}{e^{xy}}$$

$$xy' = \frac{12}{e^{xy}} - y$$

$$y' = \frac{\frac{12}{e^{xy}} - y}{x}$$

Factoring is not necessary. Divide by x.

g) $\ln xy = 10x$ — Copy the expression.

$\frac{1}{xy}(1 \cdot y + x \cdot y') = 10$ — Differentiate, using the Chain Rule for $\ln x$. Remember the Product Rule for the xy-term. The derivative of $10x$ is 10.

$\frac{1}{xy}(y + xy') = 10$ — Isolate by multiplying by xy.

$y + xy' = 10xy$ — Subtract y.

$$xy' = 10xy - y$$

$$y' = \frac{10xy - y}{x}$$

Factoring is not necessary. Divide by x for the final answer.

h) $x^2 + y^2 = 16$ — Copy the expression.

$2x + 2yy' = 0$ — Differentiate, using the Chain Rule on y^2. The derivatives of x^2 and 16 are $2x$ and 0, respectively.

$2yy' = -2x$ — Isolate the y'-term by subtracting $2x$.

$y' = \frac{-2x}{2y} = -\frac{x}{y}$ — Divide by $2y$ and simplify.

We could have solved example 3.7(h) through conventional methods by solving the original equation for y. The result is

$$y = \sqrt{16 - x^2} = (16 - x^2)^{1/2}$$

If we use the Chain Rule on this function, the result is

$$y' = \frac{1}{2}(16 - x^2)^{-1/2}(-2x)$$

$$= -x(16 - x^2)^{-1/2} \qquad \textbf{Note: } \left(\frac{1}{2}\right)(-2x) = -x$$

$$= -\frac{x}{(16 - x^2)^{1/2}}$$

Since we have started with the function $y = \left(16 - x^2\right)^{1/2}$, the end result becomes $y' = -x/y$. This result is the same found through implicit differentiation.

Two final notes of interest.

1. Be careful that you don't confuse y' (y-prime) with y^1 (y to the first power).
2. Some texts use $\frac{dy}{dx}$ (Leibniz Notation) instead of y'. Some students prefer this substitution also.

3.9 DERIVATIVES OF HIGHER ORDER

The differentiation process can be extended from the 1st derivative to the 2nd, 3rd, etc., by taking the derivative of each successive derivative.

The application is very straightforward when we treat each new derivative as an original function. The notation for higher order derivatives is shown in Table 3.3.

Original Function	y	y	$f(x)$
1st Derivative	y'	$\frac{dy}{dx}$	$f'(x)$
2nd Derivative	y''	$\frac{d^2y}{dx^2}$	$f''(x)$
3rd Derivative	y'''	$\frac{d^3y}{dx^3}$	$f'''(x)$
4th Derivative	y''''	$\frac{d^4y}{dx^4}$	$f''''(x)$

Table 3.3 Higher Order Derivatives

EXAMPLE 3.8

Find the 3rd derivative of the following functions.

a) $f(x) = x^3 - 7x^2 - 2x + 1$

b) $y = \frac{2-x}{x+5}$ (Use Leibniz notation)

c) $y' = 6e^{-4x}$ (Use y-prime notation)

SOLUTION 3.8

a) $f(x) = x^3 - 7x^2 - 2x + 1$ Copy the function.

$f'(x) = 3x^2 - 14x - 2$	Find the 1st derivative using the Power Rule.
$f''(x) = 6x - 14$	Find the 2nd derivative.
$f'''(x) = 6$	Find the 3rd derivative.

Each successive derivative acted upon the previous derivative.

b) $y = \dfrac{2-x}{x+5}$ — Copy the function.

$$\frac{dy}{dx} = \frac{(-1)(x+5) - (2-x)(1)}{(x+5)^2}$$

Use the Quotient Rule to find the first derivative.

$$\frac{dy}{dx} = \frac{-x-5-2+x}{(x+5)^2} = \frac{-7}{(x+5)^2}$$

$$\frac{d^2y}{dx^2} = \frac{(0)(x+5)^2 - (-7)[2(x+5)^1]}{[(x+5)^2]^2}$$

Find the 2nd derivative using the Quotient Rule.

$$\frac{d^2y}{dx^2} = \frac{0+7(2)(x+5)}{(x+5)^4}$$

$$\frac{d^2y}{dx^2} = \frac{14(x+5)}{(x+5)^4} = \frac{14}{(x+5)^3}$$

$$\frac{d^3y}{dx^3} = \frac{(0)(x+5)^3 - 14[3(x+5)^2(1)]}{[(x+5)^3]^2}$$

Find the 3rd derivative using the Quotient Rule.

$$\frac{d^3y}{dx^3} = \frac{0-14(3)(x+5)^2}{(x+5)^6}$$

$$\frac{d^3y}{dx^3} = \frac{-42(x+5)^2}{(x+5)^6} = -\frac{42}{(x+5)^4}$$

Note that the power of the denominator rises by one for each successive derivative.

c) $y = 6e^{-4x}$ — Copy the function.

$y' = 6e^{-4x}(-4)$ — Find the 1st derivative using the Chain Rule for e^x.

$y' = -24e^{-4x}$

$y'' = -24e^{-4x}(-4)$ — Find the 2nd derivative using the Chain Rule for e^x.

$y'' = 96e^{-4x}$

$y''' = 96e^{-4x}(-4)$ — Find the 3rd derivative using the Chain Rule for e^x.

$y''' = -384e^{-4x}$

Practice Exercises

1. For the following functions use the formula

$$m_{\tan} = \lim_{h \to 0} \frac{f(x+h) - f(x)}{h}$$

to find the derivative.

a) $f(x) = 7x + 2$

b) $f(x) = 3x^2 - x$

2. Find the derivative of the following functions.

a) $f(x) = 3x - 5$

b) $f(x) = 6x^2 - 5x + 72$

c) $f(x) = \sqrt{x^3}$

d) $f(x) = 5\sqrt[3]{x} - 2x^{-2}$

e) $f(x) = x^2$

f) $f(x) = 1 + \sqrt{2}$

3. Use the General Power Rule to find the derivative of the following functions.

a) $f(x) = (2x - 1)^6$

b) $f(x) = 3(x^2 + 7x)^4$

c) $f(x) = \dfrac{1}{(x-1)^2}$

d) $f(x) = \dfrac{8}{\sqrt[3]{(x+1)^2}}$

4. Find the derivative of the following functions using the product and quotient rules.

a) $f(x) = (x^2 - 3)(5x^3 + 2x)$

b) $f(x) = \dfrac{(2x-5)}{(7x+3)}$

c) $f(x) = (x-1)^7 (4x^2 - x)^5$

d) $f(x) = \left(\dfrac{4x-3}{x^2+5}\right)^2$

e) $f(x) = (3x-2)^{1/2} (x+6)^{1/4}$

f) $f(x) = \sqrt{\dfrac{6x+1}{5x-3}}$

5. Find the derivative of the following functions.

a) $f(x) = \dfrac{1}{2}e^x$

b) $f(x) = 2x + 4\ln x$

c) $f(x) = e^{x^2 - 5}$

d) $f(x) = \ln(x^3 - 2x^2 - 6)$

e) $f(x) = \ln\left(\dfrac{e^x}{x}\right)$

f) $f(x) = e^{6x + \ln x^2}$

6. Find the following derivatives through implicit differentiation.

   a) $4y - xy = 6$

   b) $\frac{x}{y} = 12x^2$

   c) $\sqrt{y} + \sqrt{x} = 5$

   d) $\ln(xy) = x^2 - 1$

   e) $e^{\sqrt{y}} = x$

   f) $x^3y + y^2 = 14$

7. Find the 3rd derivative of the following functions.

   a) $f(x) = x^4 - 2x^3 + 3x^2 - x$

   b) $f(x) = \frac{x}{x-1}$

   c) $y = 4\ln x^2$

Answers

1. a) 7

   b) $6x-1$

2. a) 3

   b) $12x-5$

   c) $\frac{3}{2}x^{1/2}$

   d) $\frac{5}{3}x^{-2/3}+4x^{-3}$

   e) $2x$

   f) 0

3. a) $12(2x-1)^5$

   b) $12(x^2+7x)^3(2x+7)$

   c) $\frac{-2}{(x-1)^3}$

   d) $\frac{-16}{3(x+1)^{5/3}}$

4. a) $25x^4-39x^2-6$

   b) $\frac{41}{(7x+3)^2}$

   c) $(x-1)^6(4x^2-x)^4(68x^2-52x+5)$

   d) $\frac{-4(4x-3)(2x^2-3x-10)}{(x^2+5)^3}$

   e) $\frac{9x+34}{4(x+6)^{3/4}(3x-2)^{1/2}}$

   f) $\frac{-23}{2(5x-3)^{3/2}(6x+1)^{1/2}}$

5. a) $\frac{1}{2}e^x$

   b) $\frac{2x+4}{x}$

   c) $2xe^{x^2-5}$

   d) $\frac{3x^2-4x}{x^3-2x^2-6}$

   e) $\frac{x-1}{x}$

   f) $\left(6+\frac{2}{x}\right)(e^{6x+\ln x^2})$

6. a) $\frac{y}{4-x}$

   b) $\frac{y-24xy^2}{x}$

   c) $\frac{-\sqrt{y}}{\sqrt{x}}$

   d) $\frac{2x^2y-y}{x}$

   e) $\frac{2\sqrt{y}}{e^{\sqrt{y}}}$

   f) $\frac{-3x^2y}{x^3+2y}$

7. a) $24x-12$

   b) $\frac{-6}{(x-1)^4}$

   c) $\frac{16}{x^3}$

4

Graphing with Derivatives

4.1 INTRODUCTION

The graph of a function and the derivative of a function are intertwined. Either of these aspects of a function can reveal interesting insights into the other.

We will first look at the lines tangent to a curve. This will be followed by a discussion about the information provided by the first and second derivatives about the graph of a function.

The subjects we will discuss in this chapter include:

a) Lines tangent to the graph of a function
b) Tangents to expressions in x and y
c) Graphs and critical points
d) Maxima and minima
e) Concavity and points of inflection
f) Tying it all together

4.2 LINES TANGENT TO A FUNCTION

When a straight line touches a curve at only one point, that line is called a tangent. Figure 4.1 shows a curve $f(x)$ with a line tangent to $f(x)$ at a point P.

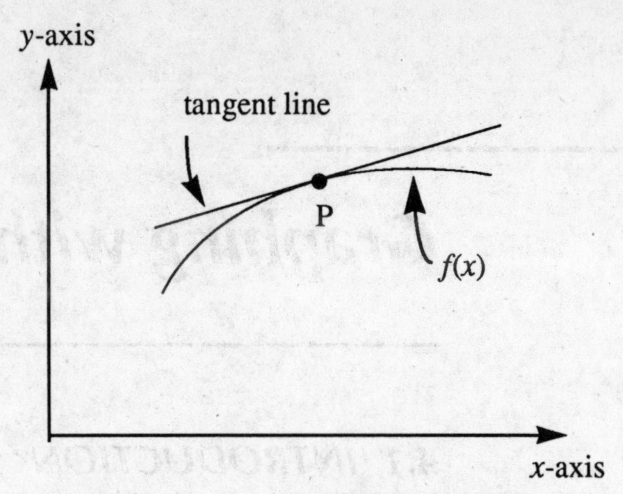

Figure 4.1 Curve with Tangent Line

We can use the first derivative of a function and the coordinates of point P to find the equation of the tangent line. The values obtained through the derivative and the point are placed into the point-slope form of the equation for a line.

> **4.1 Point-Slope**
>
> $$y - y_1 = m(x - x_1)$$
>
> where m = slope and (x_1, y_1) = point P.

There are two other formulas that help us with graphing a line. One formula is the slope formula, while the other is the slope-intercept formula.

> **4.2 Slope**
>
> $$m = \frac{y_2 - y_1}{x_2 - x_1}$$
>
> for two points (x_1, y_1) and (x_2, y_2).

4.3 Slope-Intercept

$y = mx + b$

where m = slope and y-intercept is $(0, b)$.

The graph of a line can be established by extending the line through two points. In Figure 4.2, a line is drawn through points (x_1, y_1) and (x_2, y_2). The line crosses the y-axis at b and the slope is the ratio of $y_2 - y_1$ to $x_2 - x_1$.

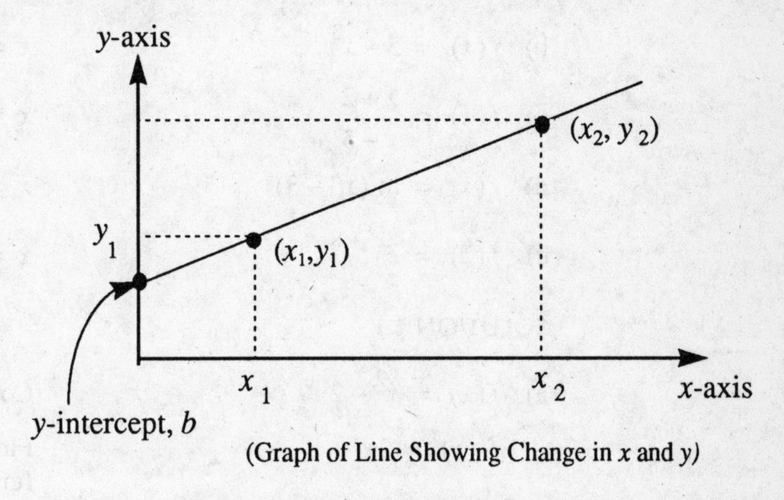

(Graph of Line Showing Change in x and y)

Figure 4.2 Typical Line Graph

Normally, we would use the points to obtain the slope. However, since the first derivative evaluated at a certain point on a function is the slope, we need only *one point* value. That slope and that point are used with the point-slope formula to find the tangent equation.

Let the x-value at a point of tangency be the constant, a. The equation for the tangent line at $x = a$ is found through the following process.

1. A function, $f(x)$, is given – along with the x-value at the point of tangency, x_1.
2. The derivative, $f'(x)$, is found.
3. The derivative is evaluated at $x = a$. In other words, find $f'(a)$ to obtain the slope, m.
4. The original function is evaluated at $x = a$ to obtain the value of y'. That is, $y_1 = f(a)$, and $x_1 = a$.

5. The values for x_1, y_1, and m are placed into the point–slope formula $y - y_1 = m(x - x_1)$.
6. The equation of the line can then be simplified.

The following examples should help to illustrate the process.

EXAMPLE 4.1

Find the line tangent to each function at the x-value indicated.

Function	At
a) $f(x) = x^2 - 2$	$x = 3$
b) $f(x) = 5 - x^3$	$x = 1$
c) $f(x) = \dfrac{x+2}{x-5}$	$x = 4$
d) $f(x) = \ln(x^2 - 3)$	$x = 2$
e) $f(x) = e^{x^3 + 8}$	$x = -2$

SOLUTION 4.1

a) $f(x) = x^2 - 2$ — Copy the given function.

$f'(x) = 2x$ — Find the derivative of the function.

$f'(3) = 2(3) = 6$ — Evaluate the derivative at $x = 3$.

$m = 6$ — The slope is 6.

$f(3) = 3^2 - 2 = 9 - 2 = 7$ — Evaluate the original function at $x = 3$.

$y_1 = 7$ — $y_1 = 7$.

$y - 7 = 6(x - 3)$
$y - 7 = 6x - 18$
$y = 6x - 11$ — Substitute $x_1 = 3$, $y_1 = 7$, $m = 6$ into the point-slope form of the equation.

b) $f(x) = 5 - x^3$ — Copy the given function.

$f'(x) = -3x^2$ — Find the derivative of the function.

$f'(1) = -3(1)^2 = -3(1) = -3$
$m = -3$ — Evaluate the derivative at $x = 1$. The slope $= -3$.

$f(1) = 5 - 1^3 = 5 - 1 = 4$	Evaluate the original function at $x = 1$.
$y_1 = 4$	$y_1 = 4$
$y - 4 = -3(x - 1)$ $y - 4 = -3x + 3$ $y = -3x + 7$	Substitute $x_1 = 1, m = -3$, and $y_1 = 4$ into the point-slope formula.
$y = -3x + 7$	The equation for the tangent line at $x = 1$ is found.
c) $f(x) = \dfrac{x+2}{x-5}$	Copy the given function.
$f'(x) = \dfrac{[1](x-5) - [1](x+2)}{(x-5)^2}$	Find the derivative.
$= \dfrac{x-5-x-2}{(x-5)^2} = \dfrac{-7}{(x-5)^2}$	
$f'(4) = \dfrac{-7}{(4-5)^2} = \dfrac{-7}{(-1)^2} = --7$	Evaluate the first derivative at $x = 4$.
$m = -7$	The slope $= -7$.
$f(4) = \dfrac{4+2}{4-5} = \dfrac{6}{-1} = -6$	Substitute $x = 4$ into the original function.
$y_1 = -6$	$y_1 = -6$
$y - (-6) = -7(x - 4)$ $y + 6 = -7x + 28$	Substitute $x_1 = 4, m = -7$, and $y_1 = -6$ into the point-slope form of the equation.
$y = -7x + 22$	The equation for the tangent line at $x = 4$ is found.
d) $f(x) = \ln(x^2 - 3)$	Copy the given function.
$f'(x) = \dfrac{1}{x^2 - 3}(2x)$	The first derivative of the function is determined.
$f'(x) = \dfrac{2x}{x^2 - 3}$	
$f'(2) = \dfrac{2(2)}{2^2 - 3} = \dfrac{4}{4-3} = \dfrac{4}{1} = 4$	Evaluate the derivative at $x = 2$.

$m = 4$	The slope = 4.
$f(2) = \ln(2^2 - 3)$ $= \ln(4-3) = \ln 1 = 0$	Evaluate the original function at $x = 2$.
$y_1 = 0$	$y_1 = 0$
$y - 0 = 4(x-2)$ $y = 4x - 8$	Substitute $x_1 = 2, m = 4$, and $y_1 = 0$ into the point-slope form of the equation.
$y = 4x - 8$	The equation for the tangent at $x = 2$ is found.
e) $f(x) = e^{x^3+8}$	Copy the given function.
$f'(x) = e^{x^3+8}(3x^2)$	Find the derivative.
$f'(-2) = e^{(-2)^3+8}(3)(-2)^2$ $= e^{-8+8}(3)(4)$ $= e^0(12) = (1)(12) = 12$	Evaluate the first derivative at $x = -2$.
$m = 12$	The slope = 12.
$f(-2) = e^{(-2)^3+8} = e^{-8+8}$ $= e^0 = 1$	Evaluate the original function at $x = -2$.
$y_1 = 1$	$y_1 = 1$
$y - 1 = 12(x - (-2))$ $y - 1 = 12(x+2)$ $y - 1 = 12x + 24$	Substitute $x_1 = -2, m = 12$, and $y_1 = 1$ into the point-slope formula.
$y = 12x + 25$	The equation of the tangent line is found.

4.3 TANGENTS TO EXPRESSIONS IN x AND y

The process of finding tangents to curves which must be implicitly differentiated is very similar to the process discussed in the previous section. In this case, both x_1 and y_1 are given. We use *both* x and y (in most cases) to find the slope.

EXAMPLE 4.2

Find the line tangent to the given point for the following expression.

a) $xy = 6$ (2, 3)

b) $xy^2 - 3y = -2$ (1, 2)

c) $1 - y^2 = \ln(2y - x)$ (1, 1)

d) $xe^y = 6$ (6, 0)

SOLUTION 4.2

a) $xy = 6$ — Copy the expression.

$$xy' + 1(y) = 0$$

Find the derivative.

$$xy' + y = 0$$
$$xy' = -y$$
$$y' = -\frac{y}{x}$$
$$y' = -\frac{3}{2} = m$$

Substitute $x_1 = 2, y_1 = 3$ into the derivative to find the slope. The slope $= -3/2$.

$$y - 3 = -\frac{3}{2}(x - 2)$$
$$y - 3 = -\frac{3}{2}x + 3$$

Place $x_1 = 2, y_1 = 3$ and $m = -3/2$ in the point-slope formula.

$$y = -\frac{3}{2}x + 6$$

The tangent line equation is found.

b) $xy^2 - 3y = -2$ — Copy the expression.

$$x(2yy') + (1)y^2 - 3y' = 0$$

Find the derivative of the expression.

$$2xyy' + y^2 - 3y' = 0$$
$$2xyy' - 3y' = -y^2$$
$$y'(2xy - 3) = -y^2$$
$$y' = \frac{-y^2}{2xy - 3}$$

$$y' = \frac{-(2)^2}{2(1)(2)-3} = \frac{-4}{4-3} = -4$$
$$m = -4$$

Substitute $x_1 = 1$ and $y_1 = 2$ into the derivative to find the slope.

$$y-2 = -4(x-1)$$
$$y-2 = -4x+4$$
$$y = -4x+6$$

Use $m = -4$, $y_1 = 2$, $x_1 = 1$ to find the equation for the tangent.

c) $1-y^2 = \ln(2y-x)$

Copy the expression.

$$-2yy' = \frac{1}{2y-x}(2y'-1)$$

Find the derivative.

$$-2yy' = \frac{2y'-1}{2y-x} = \frac{2y'}{2y-x} - \frac{1}{2y-x}$$

$$-2yy' - \frac{2y'}{2y-x} = -\frac{1}{2y-x}$$

$$-2yy' - \frac{2y'}{2y-x} = -\frac{1}{2y-x}$$

Move all terms *not* containing y'.

$$2yy' + \frac{2y'}{2y-x} = \frac{1}{2y-x}$$

Multiply both sides by –1.

$$y'\left(2y + \frac{2}{2y-x}\right) = \frac{1}{2y-x}$$

Factor out y'.

$$y'\frac{(4y^2-2xy+2)}{(2y-x)} = \frac{1}{(2y-x)}$$

Use $2y-x$ as the common denominator.

$$y'(4y^2-2xy+2) = 1$$

Multiply both sides of the equation by $2y-x$.

$$y' = \frac{1}{4y^2-2xy+2}$$

Divide by $4y^2-2xy+2$.

$$y' = \frac{1}{4(1)^2-2(1)(1)+2}$$
$$= \frac{1}{4-2+2} = \frac{1}{4}$$
$$m = \frac{1}{4}$$

Substitute $x_1 = 1$ and $y_1 = 1$ into the derivative to find the slope.

$$y-1 = \frac{1}{4}(x-1)$$

$$y-1 = \frac{1}{4}x - \frac{1}{4}$$

Use $m = \frac{1}{4}$, $y_1 = 1$, $x_1 = 1$ to find the equation of the tangent.

$$y = \frac{1}{4}x + \frac{3}{4}$$

d) $xe^y = 6$ — Copy the expression.

$xe^y y' + e^y = 0$ — Find the derivative.

$$xe^y y' = -e^y$$

$$y' = \frac{-e^y}{xe^y} = -\frac{1}{x}$$

$$y' = -\frac{1}{6}$$

Substitute $x_1 = 6$ into the derivative to find the slope.

$$m = -\frac{1}{6}$$

$$y - 0 = -\frac{1}{6}(x-6)$$

$$y = -\frac{1}{6}x + 1$$

Use $x_1 = 6$, $y_1 = 0$, and $m = -1/6$ to find the equation of the tangent.

4.4 GRAPHS AND CRITICAL POINTS

As we have seen, derivatives are used to determine the equations of lines tangent to a curve. Derivatives of a function can also be used to give us information about the function itself. This information takes on different meanings depending upon whether the derivative is positive or negative, or where *critical points* occur. In most Business and Life Sciences Calculus texts, *critical points* are defined as points on the x-axis where the derivative is zero or where the derivative is not continuous.

This section will explore the techniques for determining critical points. The interval between these critical points can be positive or negative, depending upon whether the function is above or below the x-axis. The region where the function curve is *above* the x-axis is *positive*. The region where the function curve is *below* the x-axis is *negative*.

The next example outlines the technique for determining critical points and finding the signs of the regions between those points.

EXAMPLE 4.3

Determine where $f(x) = 0$ in the following functions. Determine the regions of the number line where the function is positive and/or where the function is negative.

a) $f(x) = x^3 + x^2 - 6x$

b) $f(x) = \dfrac{x+3}{x-5}$

c) $f(x) = x^{2/3} - 1$

d) $f(x) = \sqrt{x-2}$

e) $f(x) = \dfrac{2}{x-3}$

SOLUTION 4.3

a) The process of determining the critical points occurs in two parts:

1. Set $f(x) = 0$ and solve for x. (Any values that result are placed where the function crosses the x-axis.) These are called zeros.
2. Determine points where restrictions on the domain occur. (See chapter 1.)

$f(x) = x^3 + x^2 - 6x$	Copy the function.
$f(x) = 0 = x^3 + x^2 - 6x$	Set the function equal to zero.
$x(x^2 + x - 6) = 0$	Solve for x.
$x(x-2)(x+3) = 0$	Factor if necessary.
$x = 0 \quad x - 2 = 0 \quad x + 3 = 0$ $x = 2 \quad x = -3$	The zeros are $x = 0$, $x = 2$, and $x = -3$.
	Indicate the critical points on the number line.

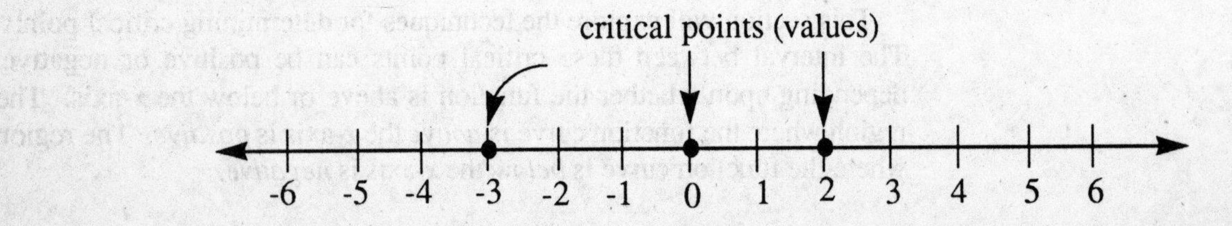

The sign of the function is determined by selecting any "test" point that occurs between the critical points. Place this "test" point into the function and observe the sign of the result.

Each line region is listed.

region A: $-\infty \to -3$

region B: $-3 \to 0$

region C: $0 \to 2$

region D: $2 \to \infty$

region	A	B	C	D
test point	–4	–1	1	3

Determine the sign of the function in each region by selecting any test point from that region.

Find the overall function sign for each test point.

$$f(x) = x^3 + x^2 - 6x$$

$$f(-4) = (-4)^3 + (-4)^2 - 6(-4) = -64 + 16 + 24$$
$$= -24 \Rightarrow -$$
$$f(-1) = (-1)^3 + (-1)^2 - 6(-1) = -1 + 1 + 6$$
$$= 6 \Rightarrow +$$
$$f(1) = (1)^3 + (1)^2 - 6(1) = 1 + 1 - 6$$
$$= -4 \Rightarrow -$$
$$f(3) = (3)^3 + (3)^2 - 6(3) = 27 + 9 - 18$$
$$= 18 \Rightarrow +$$

region	A	B	C	D
test point	–4	–1	1	3
sign	–	+	–	+

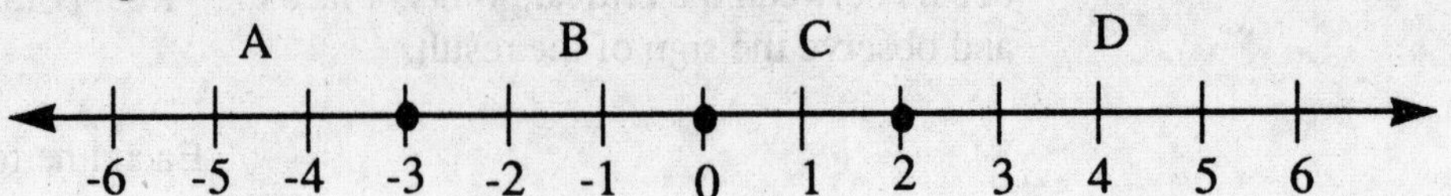

Complete the number line showing critical points and signs of regions.

Sign Test

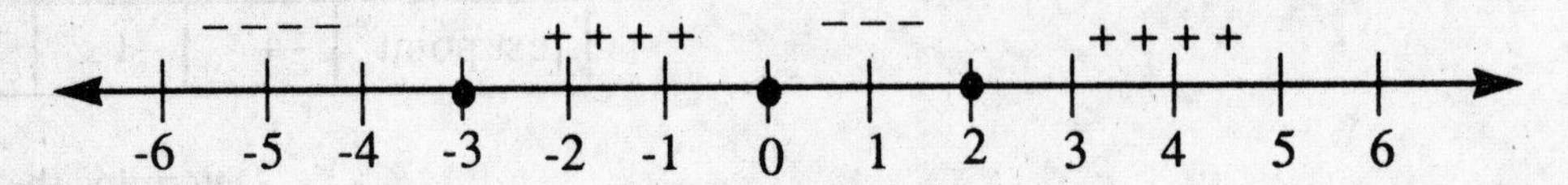

A graph of $f(x) = x^3 + x^2 - 6x$ is shown below:

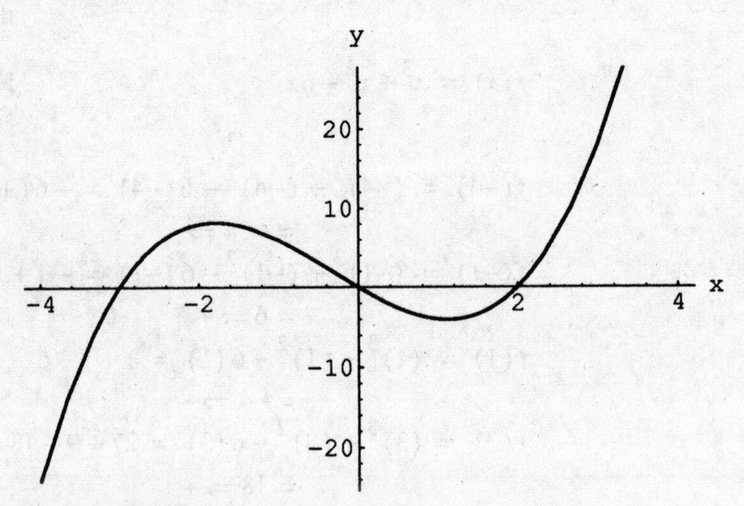

Compare this with the results of the sign test.

b) $f(x) = \dfrac{x+3}{x-5}$

Copy the given function.

A fraction is equal to zero *only* when its *numerator* equals zero.

$x + 3 = 0$
$x = -3$

The zero (–3) is found by setting $x + 3 = 0$.

$x - 5 = 0$
$x = 5$

There is a *restriction* on the domain when the *denominator* is zero.

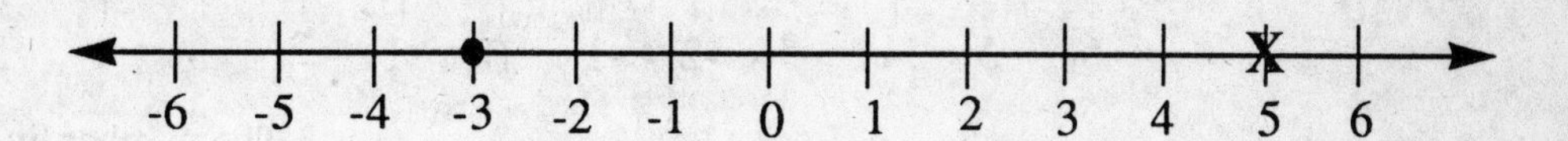

Indicate the critical points on the number line. (5 is marked with an X because it is not in the domain).

The number line is divided into *A*, *B*, and *C* regions.

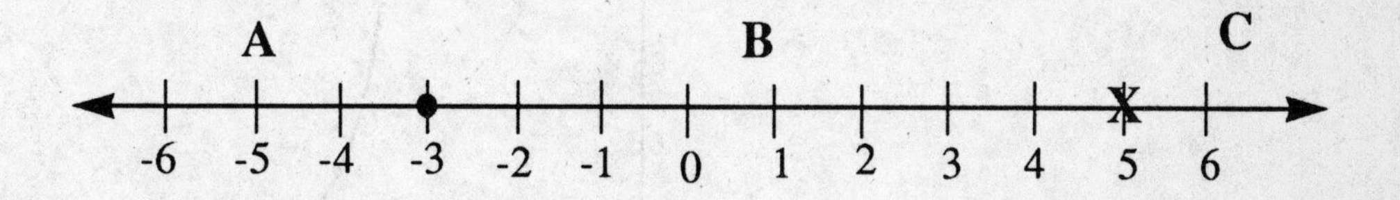

region *A*: $-\infty \to -3$
region *B*: $-3 \to 5$
region *C*: $5 \to \infty$

region	*A*	*B*	*C*
test point	–4	0	6
sign	+	–	+

A test point is selected for each region and a sign determined for that region by evaluating the function at the test point.

$$f(-4) = \frac{-4+3}{-4-5} = \frac{-1}{-9} = \frac{1}{9} \Rightarrow +$$

$$f(0) = \frac{0+3}{0-5} = -\frac{3}{5} \Rightarrow -$$

$$f(6) = \frac{6+3}{6-5} = \frac{9}{1} = +$$

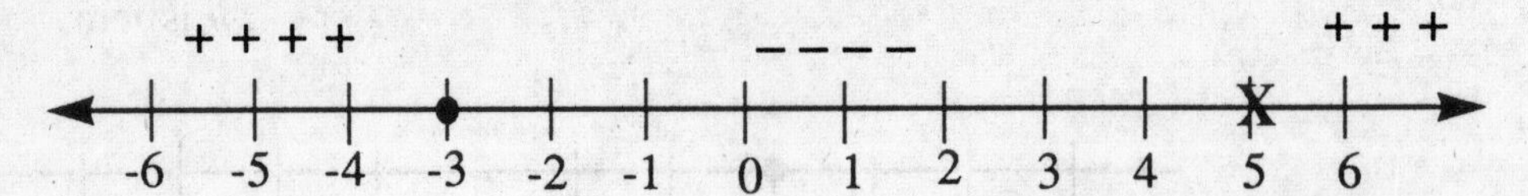

The number line is shown with the critical points (the zero and the point of discontinuity) and the sign of each region.

The graph of $f(x) = \dfrac{x+3}{x-5}$ is presented below.

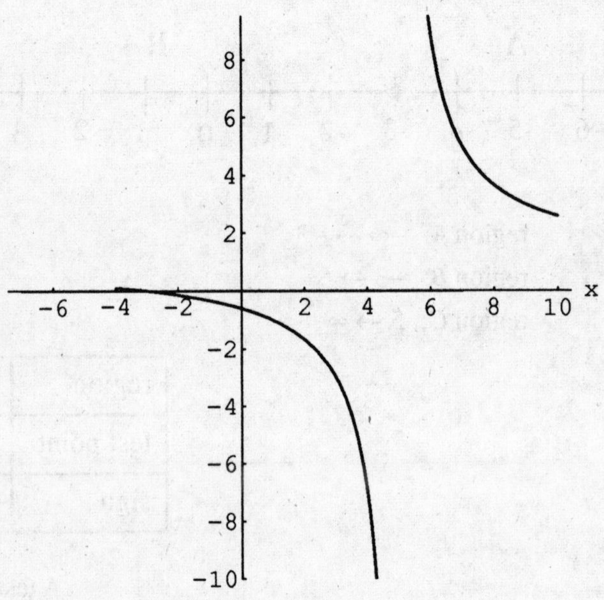

Compare with the results from the sign test.

c) $f(x) = x^{2/3} - 1$ — Copy the given function.

$x^{2/3} - 1 = 0$ — Solve for zeros.

$x^{2/3} = 1$ — Let the function equal zero and solve for the zeros.

$(x^{2/3})^3 = 1^3$ — Cube both sides.

$x^2 = 1^3 = 1$

$\sqrt{x^2} = \sqrt{1}$ — Take the square root of both sides.

$x = \pm 1$ — The critical points are –1 and 1.

Indicate the zeros on the number line.

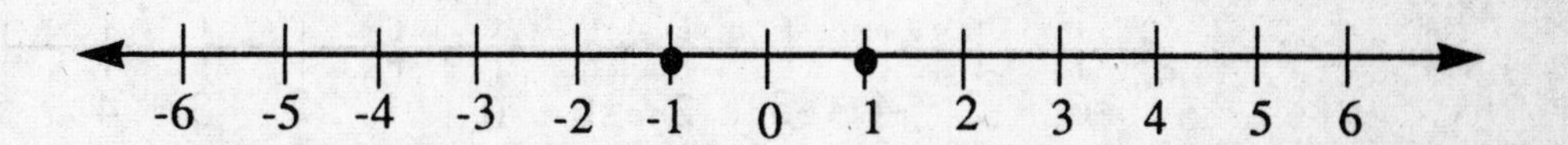

The number line is divided into regions *A*, *B*, *C*.

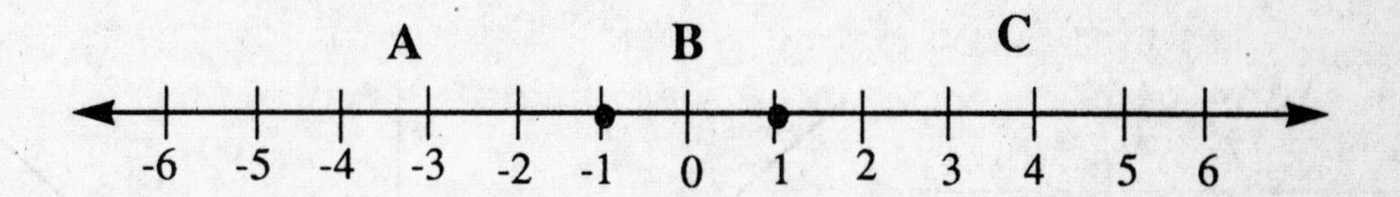

region *A*: $-\infty \rightarrow -1$
region *B*: $-1 \rightarrow 1$
region *C*: $1 \rightarrow \infty$

region	*A*	*B*	*C*
test point	–8	0	8
sign	+	–	+

A test point is selected for each region and a sign determined for that region.

$$f(-8) = (-8)^{2/3} - 1 = \sqrt[3]{(-8)^2} - 1$$

$$= \sqrt[3]{64} - 1 = 4 - 1 = 3 \Rightarrow +$$

$$f(0) = (0)^{2/3} - 1 = \sqrt[3]{0^2} - 1$$
$$= 0 - 1 = -1 \Rightarrow -$$

$$f(8) = (8)^{2/3} - 1 = \sqrt[3]{8^2} - 1$$

$$= \sqrt[3]{64} - 1 = 4 - 1 = 3 \Rightarrow +$$

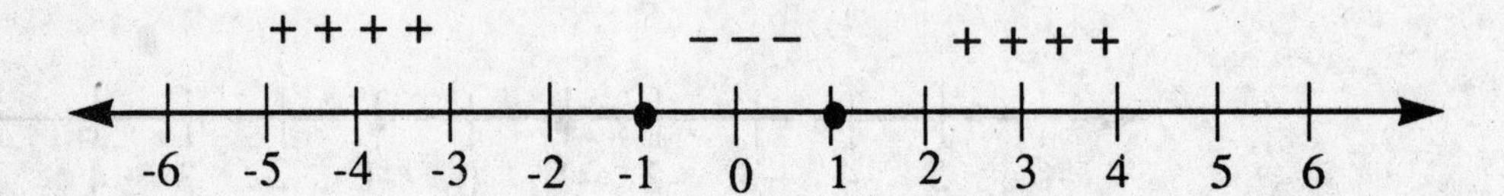

The number line is shown with the zeros and the signs for each region.

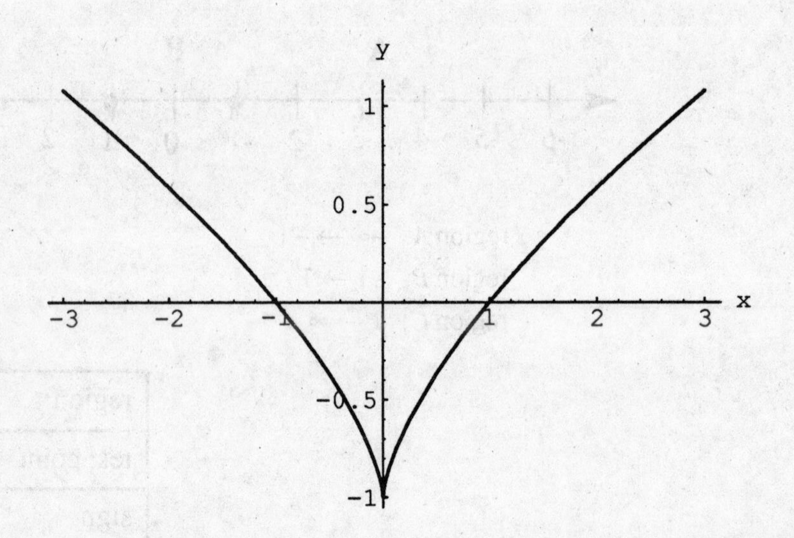

Compare the graph of $f(x) = x^{2/3} - 1$ with the results of the sign test and the critical points.

d) $f(x) = \sqrt{x-2}$ — Copy the function.

$f(x) = \sqrt{x-2} = 0$

$(\sqrt{x-2})^2 = 0^2$

$(x-2) = 0$

Determine the zeros by solving the function when it is equal to zero.

$x = 2$ — The zero is 2.

$f(x) = \sqrt{x-2} = \sqrt[2]{x-2}$

$x - 2 \geq 0$

$x \geq 2$

When radical expressions have an even index number (in this case, 2), the domain is restricted. The radicand must begreater than or equal to zero.

The function does *not* exist below 2 and equals 0 at 2.

region A: $2 \Rightarrow \infty$ — There is only one region to check. The region from 2 to infinity (region A).

$f(3) = \sqrt{3-2} = \sqrt{1} = 1 \Rightarrow +$

region	A
test point	3
sign	+

Test region A with any test point greater than 2.

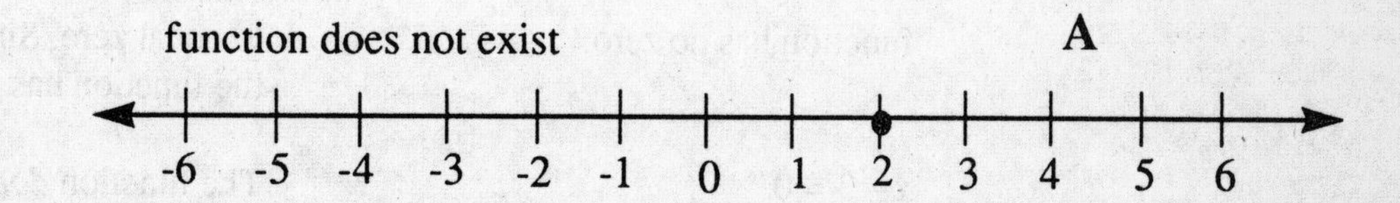

Indicate the zero and the sign of each region on the number line.

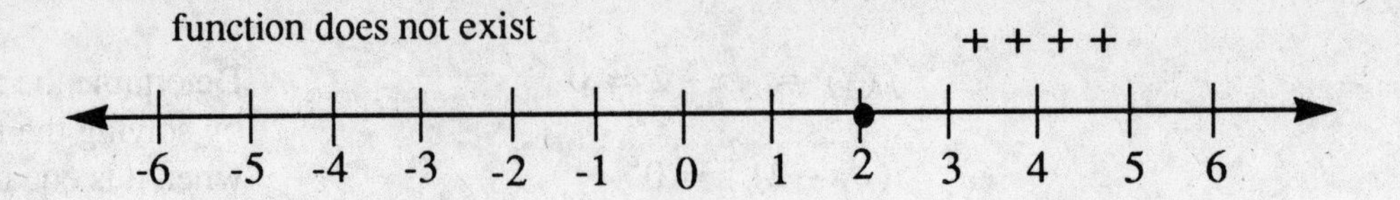

Compare the graph with the critical points and the sign test.

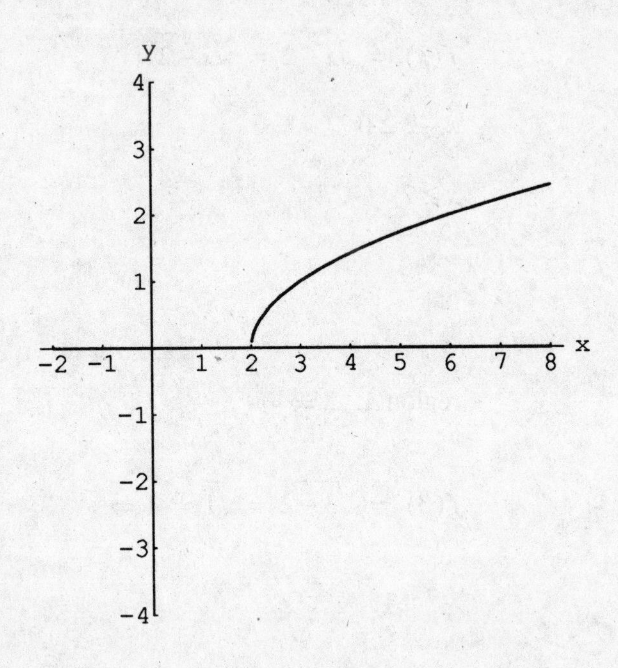

e) $f(x) = \dfrac{2}{x-3}$ — Copy the function.

$f(x) = \dfrac{2}{x-3} \quad 2 \neq 0$

function has no zero.

The numerator of a fraction must be zero for the fraction to equal zero. Since $2 \neq 0$, the function has no zero.

$x - 3 = 0$
$x = 3$

The function *does not exist* when the denominator equals zero. The function is not defined at $x = 3$.

region A: $-\infty \Rightarrow 3$
region B: $3 \Rightarrow \infty$

Indicate the point of discontinuity on the number line. The number line is divided into regions A and B. (3 is marked with an X because it is *not* part of the domain.)

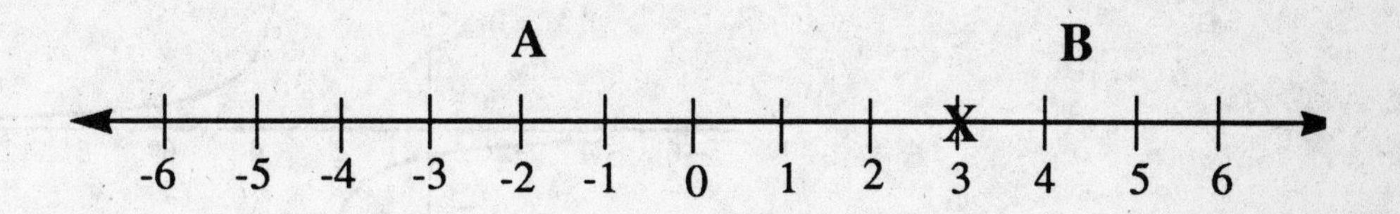

region	A	B
test point	0	4
sign	–	+

$$f(0) = \frac{2}{0-3} = -\frac{2}{3} \Rightarrow -$$

$$f(4) = \frac{2}{4-3} = \frac{2}{1} \Rightarrow +$$

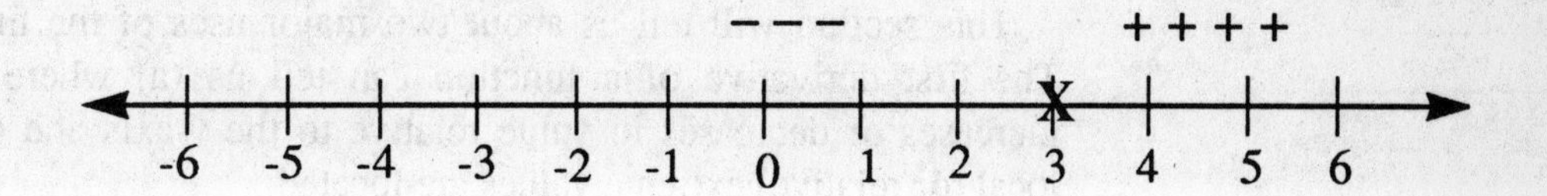

Indicate the point of discontinuity and display the appropriate sign for each region on the number line.

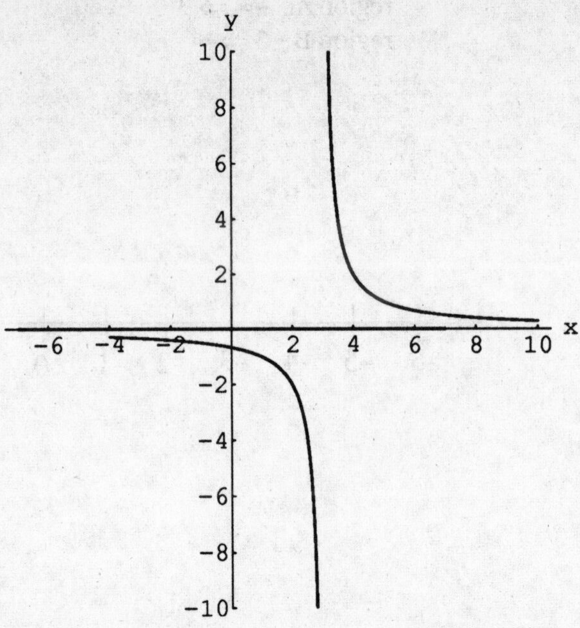

Compare the results of the sign test with the graph of $f(x) = \dfrac{2}{x-3}$.

4.5 RELATIVE EXTREMA AND THE FIRST DERIVATIVE

This section will tell us about two major uses of the first derivative. The first derivative of a function can tell us (a) where the function increases or decreases in value relative to the *y*-axis and (b) where the local (or relative) extreme values are located.

Figure 4.3 shows three lines tangent to a function, $f(x)$. Line *A* has a *positive* slope and is tangent to the portion of $f(x)$ where $f(x)$ is *increasing* (from left to right). Line *C* has a *negative* slope and it is tangent to that portion of the curve that is *decreasing* (from left to right).

The function, $f(x)$, is stationary at *one* point, P_B. Line *B* is tangent at this point and has a slope of zero. Note that point P_B is maximum *relative* to its neighboring points to the right and to the left. We can use the fact that the slope of the tangent at the extreme point is zero. We need only to know where the first derivative is zero to find the tangents of zero slope. Therefore, many "relative" extreme values for the function occur where the first derivative of a function is zero (a critical point for the first derivative). The first derivative (and all subsequent derivatives) also do not exist at a location on the number line where the original function does not exist. The most positive value for an *entire* function is called the *global maximum.* The most negative value for an *entire* function is called the *global* minimum.

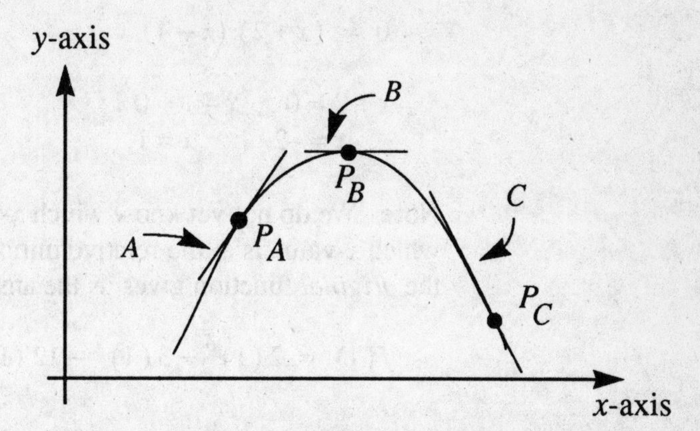

Figure 4.3 Lines Tangent to a Curve

Therefore, the first derivative can be used to determine points of discontinuity *and* extreme values. The extreme value that is greater than those around it is called a *relative maximum.* The extreme value that is less than those around it is called a *relative minimum.*

EXAMPLE 4.4

Use the first derivative of each of the following functions to determine the location of any maxima, minima, and discontinuities.

a) $f(x) = 2x^3 + 3x^2 - 12x + 1$

b) $f(x) = x^4 - 8x^2 + 3$

c) $f(x) = \dfrac{x}{(2-x)^2}$

SOLUTION 4.4

a) $f(x) = 2x^3 + 3x^2 - 12x + 1$ — Copy the function.

$f'(x) = 6x^2 + 6x - 12$ — Find the first derivative.

$= 6x^2 + 6x - 12 = 0$ — Set the first derivative equal to zero to find the relative extrema of the original function.

$$\frac{0}{6} = \frac{6x^2}{6} + \frac{6x}{6} - \frac{12}{6}$$ Divide by 6.

$$0 = x^2 + x - 2$$

$$0 = (x+2)(x-1)$$ Factor.

$$x+2=0 \quad x-1=0$$
$$x=-2 \quad x=1$$ The extrema occur at $x = -2$ and $x = 1$.

Note: We do not yet know which x-value is at the relative maximum and which x-value is at the relative minimum. Substitution of these values in the *original* function gives us the answer.

$$f(1) = 2(1)^3 + 3(1)^2 - 12(1) + 1 = 2 + 3 - 12 + 1 = -6$$

relative minimum

$$f(-2) = 2(-2)^3 + 3(-2)^2 - 12(-2) + 1 = 2(-8) + 3(4) + 24 + 1$$
$$= -16 + 12 + 24 + 1 = 21$$

relative maximum

The relative maximum occurs at (–2, 21), and (1, –6) is the relative minimum.

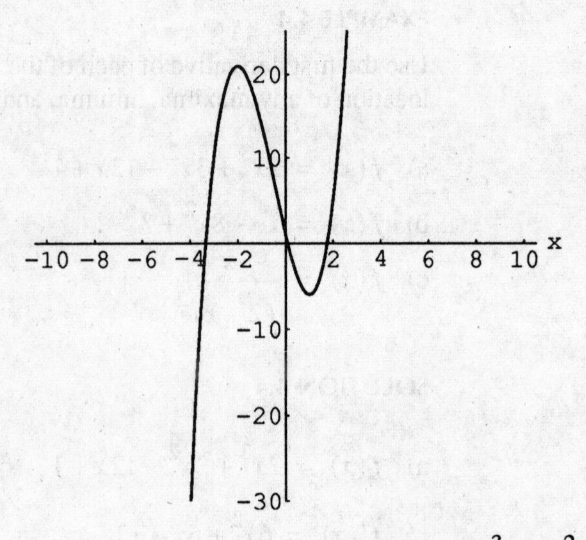

$f(x) = 2x^3 + 3x^2 - 12x + 1$ is graphed at left. As you can see, the overall (global) maximum and minimum are ∞ and $-\infty$, respectively.

b) $f(x) = x^4 - 8x^2 + 3$	Copy the function.
$f'(x) = 4x^3 - 16x$	Find the first derivative of the function.
$(4x^3 - 16x) = 0$ $4x(x^2 - 4) = 0$ $4x(x+2)(x-2) = 0$	Set the first derivative equal to zero and solve for the critical points (zeros and points of discontinuity, if any).
$4x = 0 \quad x+2=0 \quad x-2=0$ $x=0 \quad x=-2 \quad x=2$	The critical points are 0, –2, and 2.
$f(-2) = (-2)^4 - 8(-2)^2 + 3$ $= 16 - 8(4) + 3$ $= 16 - 32 + 3 = -13$ $f(0) = 0^4 - 8(0)^2 + 3$ $= 0 - 0 + 3 = 3$ $f(2) = 2^4 - 8(2)^2 + 3$ $= 16 - 8(4) + 3$ $= 16 - 32 + 3 = 19 - 32 = -13$	Find the relative maximum or relative minimum by substituting the critical points into the *original* function.
minimums (–2, –13) and (2, –13) maximum (0, 3)	The function has two relative minimums and one relative maximum.

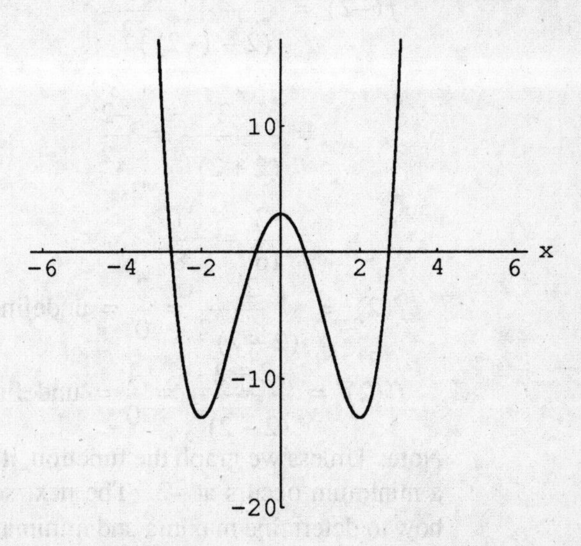

The graph of $f(x) = x^4 - 8x^2 + 3$ is drawn. The global maximum is ∞. The global minimums are also the relative minimums.

c) $f(x) = \dfrac{x}{(2-x)^2}$ — Copy the function.

$$f'(x) = \frac{[1](2-x)^2 - (x)2(2-x)(-1)}{((2-x)^2)^2}$$

Find the first derivative.

$$f'(x) = \frac{(2-x)^2 + 2x(2-x)}{(2-x)^4}$$

$$f'(x) = \frac{\cancel{(2-x)}((2-x)+2x)}{\cancel{(2-x)}(2-x)^3}$$

Factor out $(2-x)$.

$$f'(x) = \frac{(2-x)+2x}{(2-x)^3}$$

Simplify.

$$f'(x) = \frac{2+x}{(2-x)^3}$$

numerator $= 2 + x = 0$

$x = -2$

Set the numerator equal to zero to find the critical point.

$$f(-2) = \frac{-2}{(2-(-2))^2}$$

Substituting –2 into the original function gives us the extreme value.

$$= \frac{-2}{(2+2)^2} = \frac{-2}{4^2}$$

$$= \frac{-2}{16} = -\frac{1}{8}$$

$$f(2) = \frac{2}{(2-2)^2} = \frac{2}{0} = \text{undefined}$$

$$f'(2) = \frac{2+2}{(2-2)^3} = \frac{4}{0} = \text{undefined}$$

Both the original function and the first derivative are discontinuous at $x = 2$.

Note: Unless we graph the function, it is not clear whether a maximum or a minimum occurs at –2. The next section will provide information on how to determine maxima and minima without the use of a graph.

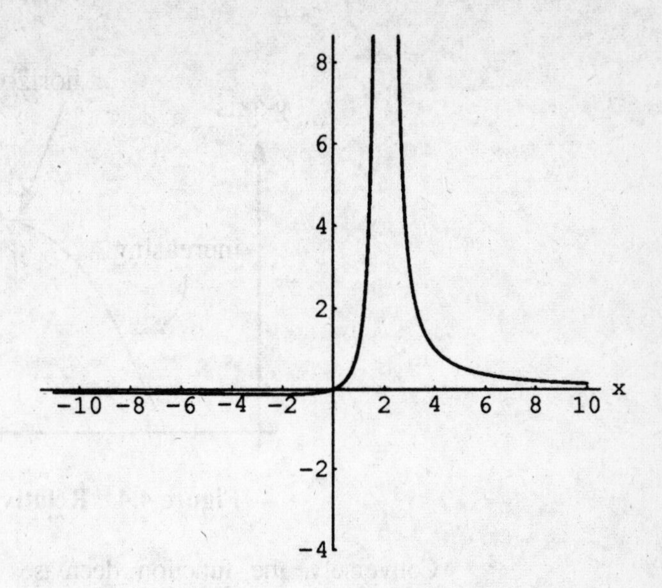

The graph of $f(x) = \dfrac{x}{(x-2)^2}$. Here we can see that $\left(-2, -\dfrac{1}{8}\right)$ is a minimum.

Regions Where a Function Increases or Decreases

At the beginning of this section, we noted that the slope of a tangent line would help determine where a function was increasing or decreasing. A tangent line with a positive slope indicates that the function is increasing. A tangent line with a negative slope occurs where the function is decreasing.

We can determine a maximum by noting that the function increases before a maximum and decreases after the maximum point is reached. See Figure 4.4.

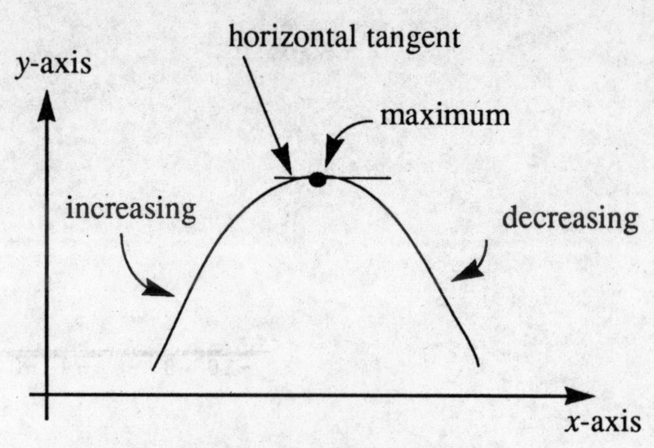

Figure 4.4 Relative Maximum

Conversely, the function decreases *before* a minimum point and increases *after*. See Figure 4.5.

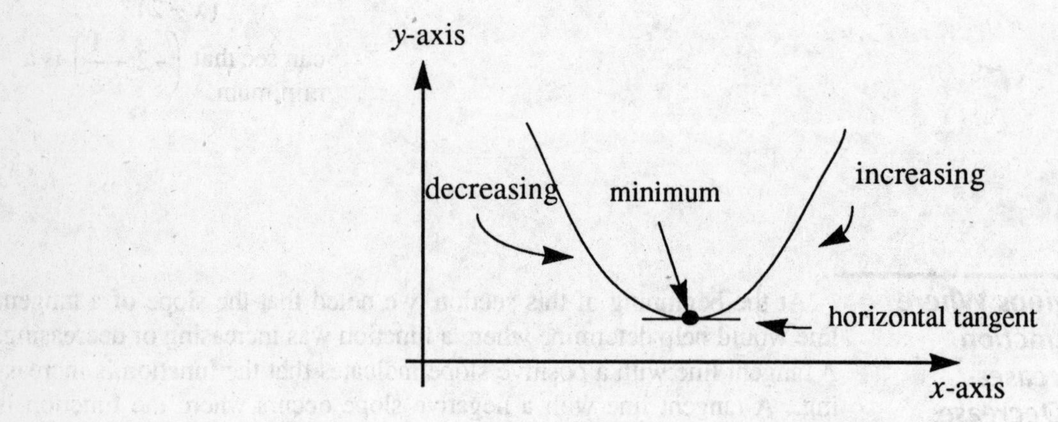

Figure 4.5 Relative Minimum

Therefore, a pattern can be set up for the examination of the signs of the first derivative. The sign test for extrema can be summarized in the following table.

Extrema	Signs for First Derivative
maximum	+ + + + + + + then – – – – – – –
minimum	– – – – – – – then + + + + + + +

Table 4.1 First Derivative Sign Test

The signs are read left to right.
Note: *Not all functions will have a relative minimum or maximum.*

EXAMPLE 4.5

Use the zeros and the signs of the first derivative to find extrema and the intervals where the function is increasing or decreasing.

a) $f(x) = x^3 - x^2 - x + 2$

b) $f(x) = 3x^2 - x^3 - 3x$

c) $f(x) = \dfrac{6}{x-1}$

SOLUTION 4.5

a) $f(x) = x^3 - x^2 - x + 2$ — Copy the function.

$f'(x) = 3x^2 - 2x - 1$ — Find the first derivative.

$3x^2 - 2x - 1 = 0$ — Set the first derivative equal to zero.

$(3x+1)(x-1) = 0$ — Factor to solve.

$3x + 1 = 0 \qquad x - 1 = 0$

$3x = -1 \qquad x = 1$

$x = -\dfrac{1}{3}$

The zeros are $-1/3$ and 1. The zeros divide the number line into three regions A, B, C.

region	A	B	C
test point	–1	0	2
sign	+	–	+

Find the sign of the first derivative for each region. Select a test point and evaluate the first derivative.

$$f'(-1) = 3(-1)^2 - 2(-1) - 1$$
$$= 3 + 2 - 1$$
$$= 4 \Rightarrow +$$
$$f'(0) = 3(0)^2 - 2(0) - 1 = -1 \Rightarrow -$$
$$f'(2) = 3(2)^2 - 2(2) - 1 = 3 \cdot 4 - 4 - 1$$
$$= 12 - 4 - 1 = 7 \Rightarrow +$$

With + indicating a region in which the function is increasing, and – a region in which the function is decreasing, detail the regions on the number line.

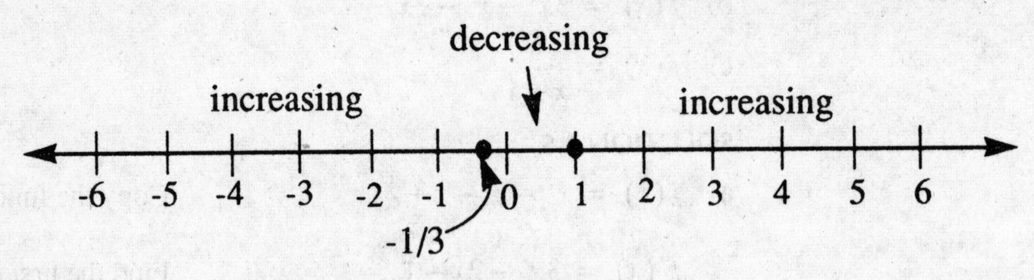

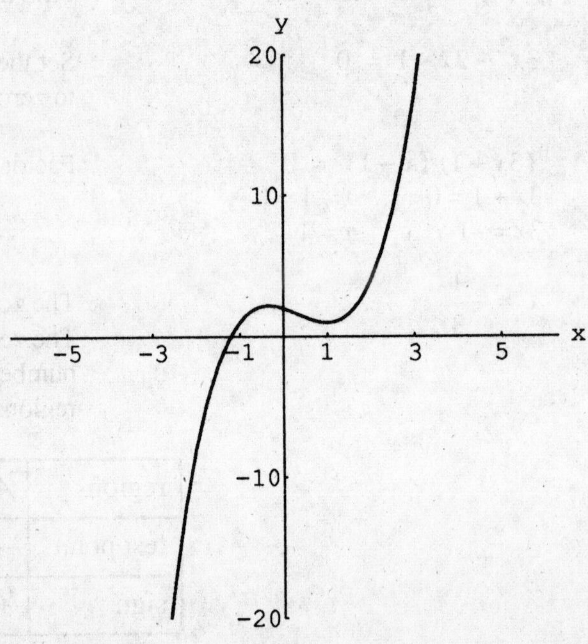

Compare $f(x) = x^3 - x^2 - x + 2$ to the information found from the first derivative.

$$f\left(-\frac{1}{3}\right) = \left(-\frac{1}{3}\right)^3 - \left(-\frac{1}{3}\right)^2 - \left(-\frac{1}{3}\right) + 2$$
$$= \frac{59}{27}$$

$$f(1) = (1)^3 - (1)^2 - (1) + 2 = 1$$

The maximum is at $\left(-\frac{1}{3}, \frac{59}{27}\right)$ and the minimum is at (1, 1).

b) $f(x) = 3x^2 - x^3 - 3x$ — Copy the original function.

$f'(x) = 6x - 3x^2 - 3$ — Find the first derivative.

$6x - 3x^2 - 3 = 0$ — Set the first derivative equal to zero.

$\frac{6x}{-3} - \frac{3x^2}{-3} - \frac{3}{-3} = \frac{0}{-3}$ — Divide by -3.

$-2x + x^2 + 1 = 0$ — Rearrange the terms.

$x^2 - 2x + 1 = 0$

$(x-1)(x-1) = 0$ — Factor.

$x - 1 = 0$

$x = 1$ — The zero is at $x = 1$.

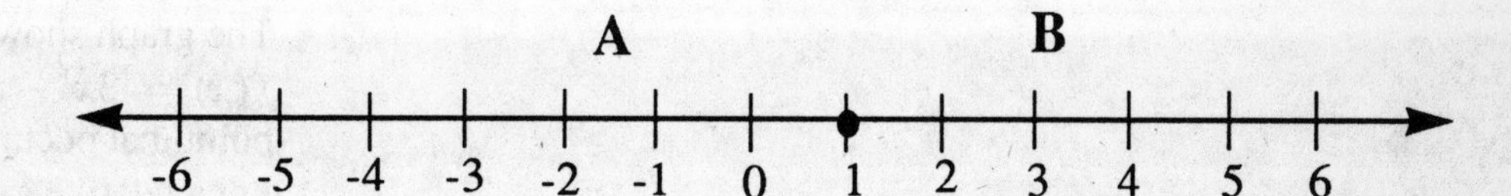

The critical point divides the number line into two regions, *A* and *B*.

region	*A*	*B*
test point	0	2
sign	–	–

Find test points for each region. Find the signs of the first derivative.

$$f'(0) = 6(0) - 3(0)^2 - 3$$
$$= -3 \Rightarrow -$$

The – signs mean that the original function is decreasing on both sides of the critical point.

$$f'(2) = 6(2) - 3(2)^2 - 3$$
$$= 12 - 3 \cdot 4 - 3 = -3 \Rightarrow -$$

Note that the critical point was found by first dividing the quadratic equation, $6x - 3x^2 - 3 = 0$ by –3. If you substitute test points into the resulting equation, $f(x) = -2x + x^2 + 1$, then the sign test will be incorrect. Always substitute test points into the *initial* first derivative function. Substitutions into the factored form may produce erroneous results.

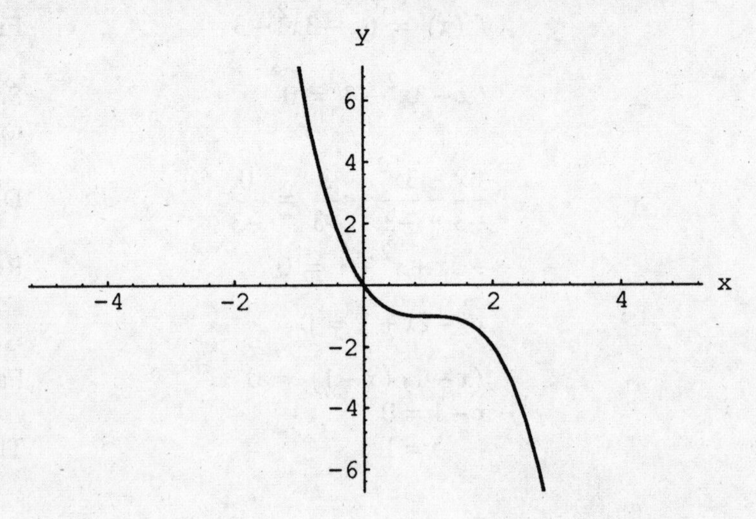

The graph shows $f(x) = 3x^2 - x^3 - 3x$. The point that occurs at $x = 1$ is *not* a relative extreme value. We will determine its nature in the next section.

$$f(1) = 3(1)^2 - 1^3 - 3(1)$$
$$= 3 - 1 - 3 = -1$$

c) $f(x) = \dfrac{6}{x-1}$

Copy the original function.

$$f'(x) = \frac{(0)(x-1) - 6(1)}{(x-1)^2}$$

Find the first derivative.

$$f'(x) = \frac{-6}{(x-1)^2}$$

The numerator of the first derivative cannot be set equal to zero, because it is a constant (–6).

$$\frac{6}{(x-1)^2} \rightarrow (x-1)^2 = 0$$

The denominator, however, gives a point of discontinuity at $x = 1$.

$$x - 1 = 0$$
$$x = 1$$

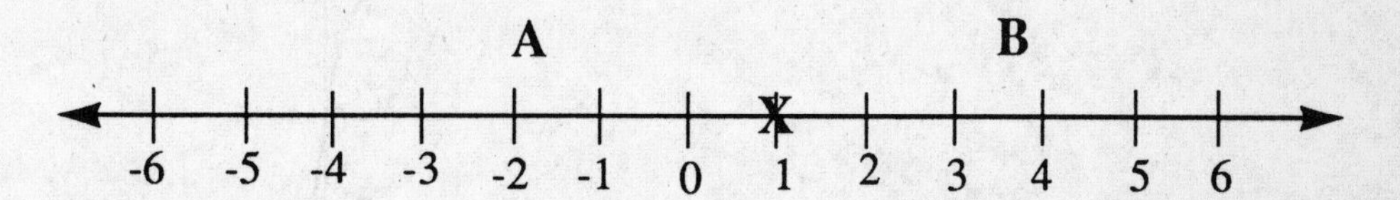

We find that the point of discontinuity divides the number line into two regions: *A* and *B*. This point is marked with an X because it is not in the domain of $f'(x)$.

region	*A*	*B*
test point	0	2
sign	–	–

$$f'(0) = \frac{-6}{(0-1)^2} = \frac{-6}{1} = -6 \Rightarrow -$$

$$f'(2) = \frac{-6}{(2-1)^2} = \frac{-6}{1} = -6 \Rightarrow -$$

Determine the sign of the first derivative at the test points. The – signs indicate that the *original* function is decreasing on both sides of the critical point.

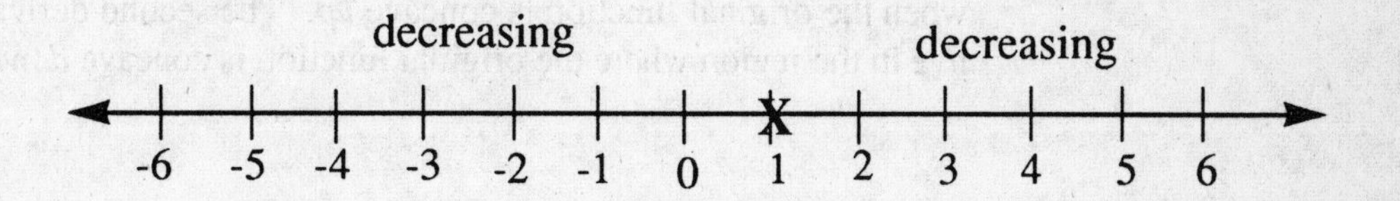

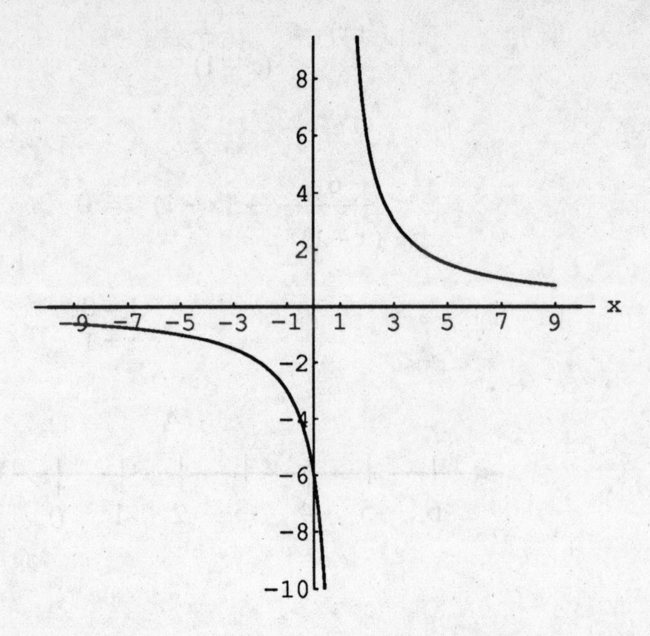

The graph of

$$f(x) = \frac{6}{(x-1)}$$

demonstrates the results of the sign test.

4.6 POINTS OF INFLECTION AND CONCAVITY

The *second derivative* of a function also plays an important role in determining the nature of the original function. The second derivative is zero where the *point of inflection* (if any) occurs. A point of inflection occurs when concavity changes *and* $f''(x) = 0$. The nature of the concave curve can be seen in Table 4.2. The second derivative is *positive* when the original function is concave *up*. The second derivative is *negative* in the region where the original function is concave *down*.

$f''(x)$ sign	Description	Shape of $f(x)$
+	concave up	
–	concave down	

Table 4.2 Signs of the Second Derivative and Concavity

The next example shows how the second derivative is used to determine the regions where the original function is concave up or concave down. A point of inflection can be found by finding the second derivative and solving for $f''(x) = 0$.

EXAMPLE 4.6

Find the regions on the graph where the following functions are concave up or concave down. Determine any points of inflection.

a) $f(x) = 4x^3 - 12x^2 + 8x - 1$

b) $f(x) = x^4 - 10x^2 + 9$

c) $f(x) = \dfrac{x+1}{x-2}$

SOLUTION 4.6

a) $f(x) = 4x^3 - 12x^2 + 8x - 1$ Copy the original function.

$f'(x) = 12x^2 - 24x + 8$ Find the first derivative.

$f''(x) = 24x - 24$ Find the second derivative.

$$(24x - 24) = 0$$
$$24x = 24$$
$$x = 1$$

Set the second derivative equal to 0 to find the critical point, which is where the point of inflection may occur.

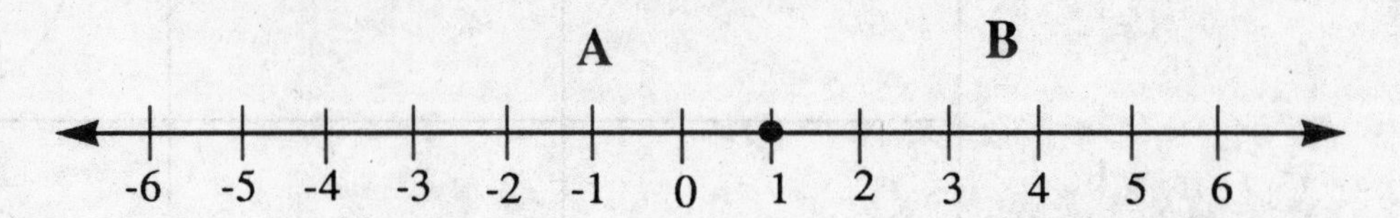

The zero divides the number line into two regions A and B.

region	A	B
test point	0	2
sign	–	+

Select test points and determine the signs of the second derivative in regions A and B.

$$f''(0) = 24(0) - 24 = -24 \Rightarrow -$$
$$f''(2) = 24(2) - 24 = 24 \Rightarrow +$$

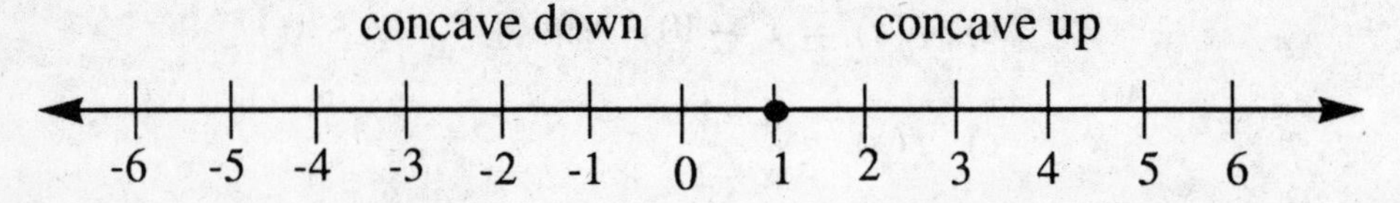

The signs of the second derivative tell us about concavity.

$$f(1) = 4(1)^3 - 12(1)^2 + 8(1) - 1$$
$$= 4 - 12 + 8 - 1 = -1$$

We find the point of inflection by evaluating the original function at $x = 1$. The point of inflection occurs at $(1, -1)$.

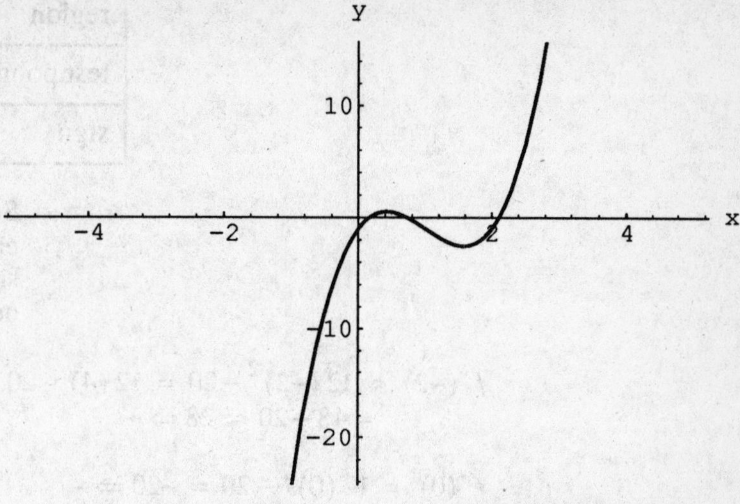

The graph of $f(x) = 4x^3 - 12x^2 + 8x - 1$ is above.

b) $f(x) = x^4 - 10x^2 + 9$ — Copy the original function.

$f'(x) = 4x^3 - 20x$ — Find the first derivative.

$f''(x) = 12x^2 - 20$ — Find the second derivative.

$(12x^2 - 20) = 0$ — Setting $f''(x) = 0$ will give us the locations of the points of inflection.

$12x^2 = 20$

$$x^2 = \frac{20}{12} = \frac{5}{3}$$

$$x = \pm\sqrt{\frac{5}{3}} \approx \pm 1.29$$

Take the square root of both sides.

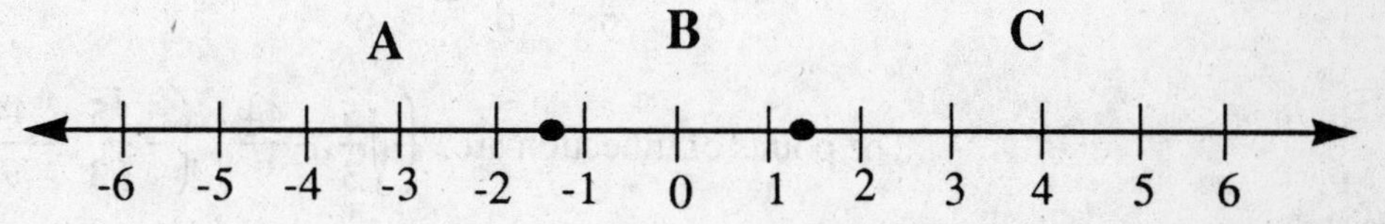

The two critical points divide the number line into three regions.

region	A	B	C
test point	–2	0	2
sign	+	–	+

Selecting a test point from each region, we can test the sign of the second derivative.

$$f''(-2) = 12(-2)^2 - 20 = 12(4) - 20$$
$$= 48 - 20 = 28 \Rightarrow +$$

$$f''(0) = 12(0)^2 - 20 = -20 \Rightarrow -$$

$$f''(2) = 12(+2)^2 - 20 = 48 - 20 = 28 \Rightarrow +$$

The signs tell us where the original function is concave up and concave down.

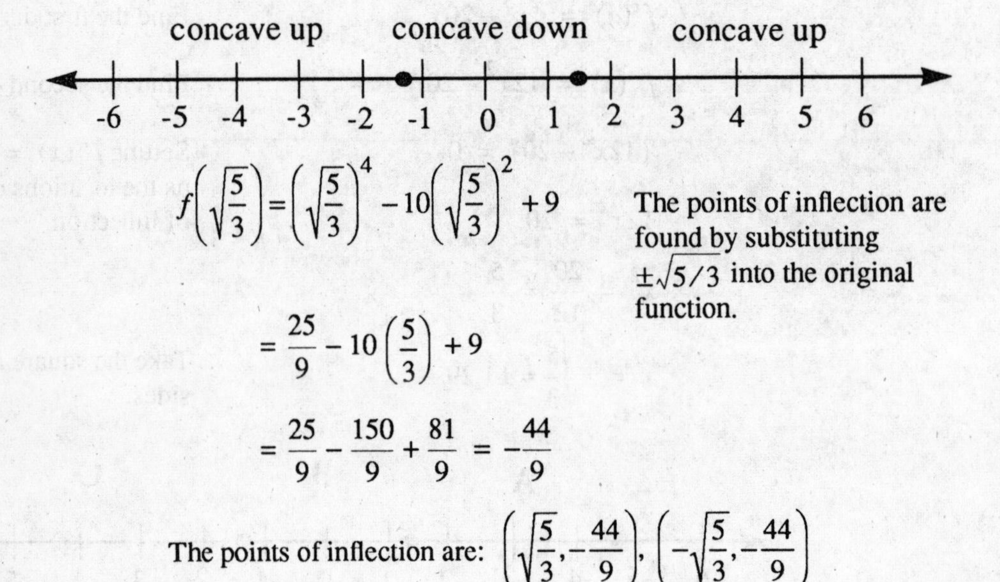

$$f\left(\sqrt{\frac{5}{3}}\right) = \left(\sqrt{\frac{5}{3}}\right)^4 - 10\left(\sqrt{\frac{5}{3}}\right)^2 + 9$$

The points of inflection are found by substituting $\pm\sqrt{5/3}$ into the original function.

$$= \frac{25}{9} - 10\left(\frac{5}{3}\right) + 9$$

$$= \frac{25}{9} - \frac{150}{9} + \frac{81}{9} = -\frac{44}{9}$$

The points of inflection are: $\left(\sqrt{\frac{5}{3}}, -\frac{44}{9}\right), \left(-\sqrt{\frac{5}{3}}, -\frac{44}{9}\right)$

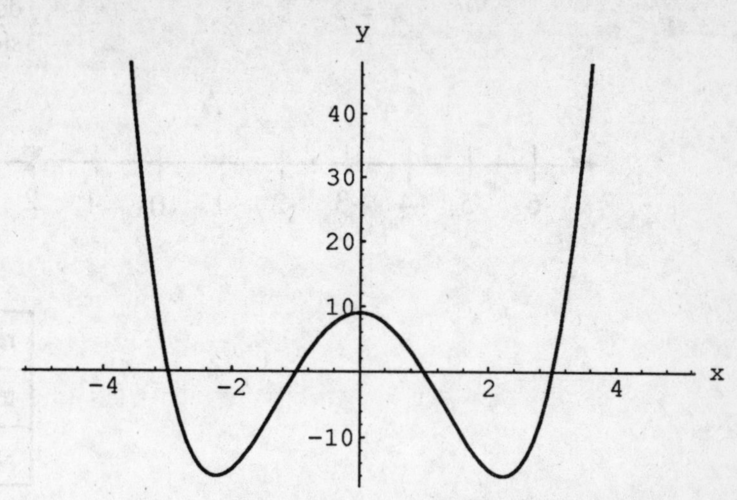

The function $f(x) = x^4 - 10x^2 + 9$ is shown above. This function has two points of inflection.

c) $f(x) = \dfrac{x+1}{x-2}$

Copy the original function.

$$f'(x) = \frac{[1]\,(x-2) - [1]\,(x+1)}{(x-2)^2}$$

Find the first derivative.

$$f'(x) = \frac{x-2-x-1}{(x-2)^2} = \frac{-3}{(x-2)^2}$$

$$f''(x) = \frac{0\,(x-2)^2 - (-3)\,[2\,(x-2)\,(1)\,]}{(x-2)^4}$$

Find the second derivative.

$$f''(x) = \frac{6\,(x-2)}{(x-2)^4}$$

$$f''(x) = \frac{6}{(x-2)^3}$$

The second derivative cannot be equal to zero because the numerator is a constant.

$$(x-2)^3 = 0$$

$$x-2 = 0$$

The point of discontinuity is where the denominator is zero. The sign of the second

$x = 2$

derivative is tested on either side of this point.

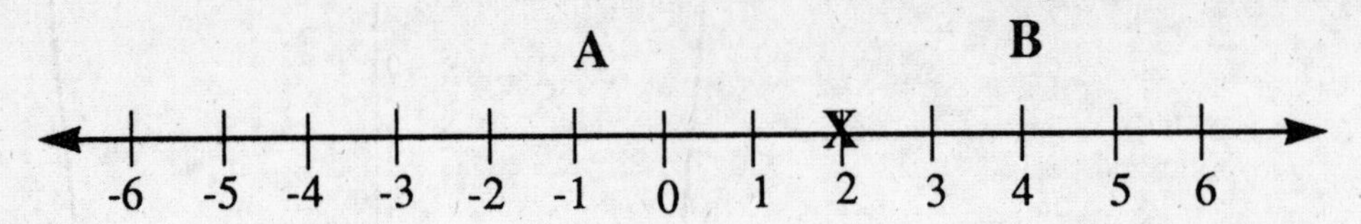

region	A	B
test point	1	3
sign	–	+

A table is made with the test points selected in regions A and B.

$$f''(1) = \frac{6}{(1-2)^3} = \frac{6}{(-1)^3} = \frac{6}{-1} \Rightarrow -$$

$$f''(3) = \frac{6}{(3-2)^3} = \frac{6}{(1)^3} = \frac{6}{1} \Rightarrow +$$

The original function is concave down on the left side of the discontinuity and concave up on the right side.

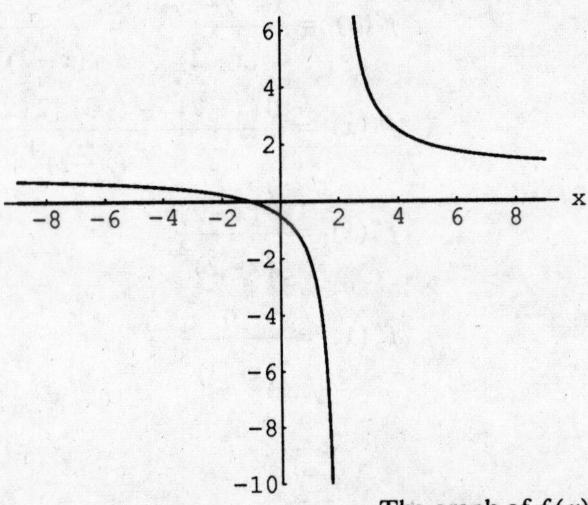

The graph of $f(x) = \dfrac{x+1}{x-2}$ is above.

4.7 TYING IT ALL TOGETHER

The combination of the information obtained from the original function, the first derivative, and the second derivative help us to determine the nature of a function's graph. The procedure used to obtain this information is outlined below.

Graphing with Derivatives

1. Given the original function.
2. Determine the y-intercept, $f(0)$.
3. Find the first derivative.
4. Set the first derivative equal to zero to find where extrema occur.
5. Use the location of the extrema to find the y-value of the extrema.
6. Use the sign test for the first derivative to find regions where the function is increasing or decreasing.
7. Find the second derivative.
8. Set the second derivative equal to zero to obtain the location of the points of inflection.
9. Place the values from the second derivative into the original function to obtain y-value of the points of inflection.
10. Use the sign test with the second derivative to find regions of concavity.

The original function is evaluated at the zeros of the first derivative and the second derivative. Students often find it confusing to return to the original function, but this confusion can be alleviated by strictly following the sequence.

EXAMPLE 4.7

Use the function and its derivatives to determine information about the graph of the following.

a) $f(x) = x^3 - 6x^2 + 9x + 4$

b) $f(x) = \dfrac{2x}{(1-x)^2}$

SOLUTION 4.7

a) $f(x) = x^3 - 6x^2 + 9x + 4$ — Copy the original function.

$f(0) = 0^3 - 6(0)^2 + 9(0) + 4$
$= 4$ — Determine the y-intercept.

$f'(x) = 3x^2 - 12x + 9$ — Find the first derivative.

$$3x^2 - 12x + 9 = 0$$

Set the first derivative equal to zero.

$$x^2 - 4x + 3 = 0$$

Divide both sides by 3.

$$(x-1)(x-3) = 0$$

$$x = 1, \ x = 3$$

The locations of the extrema are at $x = 1$ and $x = 3$.

$$f(1) = 1^3 - 6(1)^2 + 9(1) + 4$$
$$= 1 - 6 + 9 + 4 = 8$$

Evaluate the *original* function at the zeros of the first derivative.

$$f(3) = 3^3 - 6(3)^2 + 9(3) + 4$$
$$= 27 - 54 + 27 + 4$$
$$= 4$$

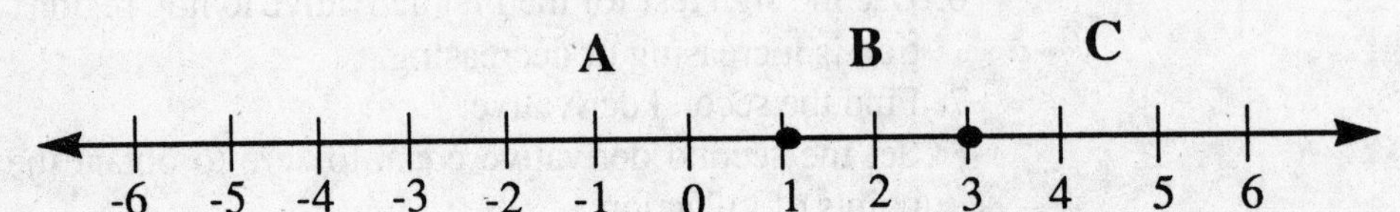

The sign test is next. The zeros of the first derivative separate the number line into three regions.

region	*A*	*B*	*C*
test point	0	2	4
sign	+	–	+

The table shows that the original function increases in region *A*, decreases in region *B*, and increases in region *C*.

(1, 8) and (3, 4) are the relative maximum and the relative minimum, respectively.

$$f'(0) = 3(0)^2 - 12(0) + 9 = 9 \Rightarrow +$$

$$f'(2) = 3(2)^2 - 12(2) + 9 = 3 \cdot 4 - 24 + 9$$
$$= 12 - 24 + 9 = -3 \Rightarrow -$$

$$f'(4) = 3(4)^2 - 12(4) + 9 = 3 \cdot 16 - 48 + 9$$
$$= 48 - 48 + 9 \Rightarrow +$$

$$f''(x) = 6x - 12$$

The second derivative is found.

$$6x - 12 = 0$$
$$6x = 12$$
$$x = 2$$

Set the second derivative equal to zero and solve for the location of the point of inflection.

$$f(2) = 2^3 - 6(2)^2 + 9(2) + 4$$
$$= 8 - 6 \cdot 4 + 18 + 4$$
$$= 8 - 24 + 18 + 4 = 6$$

The point of inflection is at (2, 6).

Substitute the zero from the second derivative in the *original* equation to determine the point of inflection.

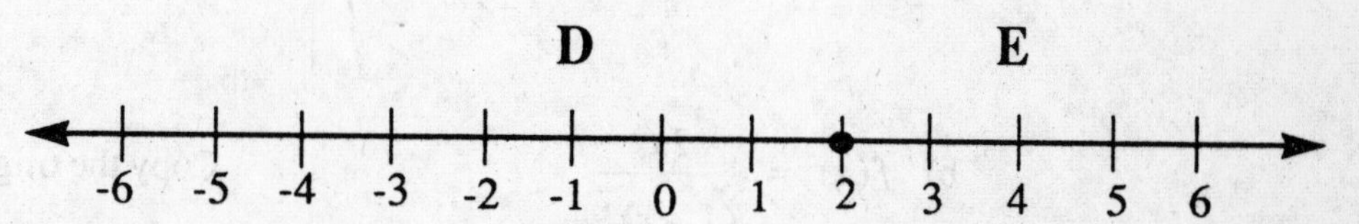

The zero for the second derivative divides the number line into two regions.

region	*D*	*E*
test point	0	3
sign	–	+

The sign test is next.

$$f''(0) = 6(0) - 12 = -12 \Rightarrow -$$
$$f''(3) = 6(3) - 12 = 6 \Rightarrow -$$

Determine the signs of the regions. The second derivative tells us that the original function is concave down in region *D* and concave up in region *E*.

The graph of the function follows. Compare the results of the procedure with the graph.

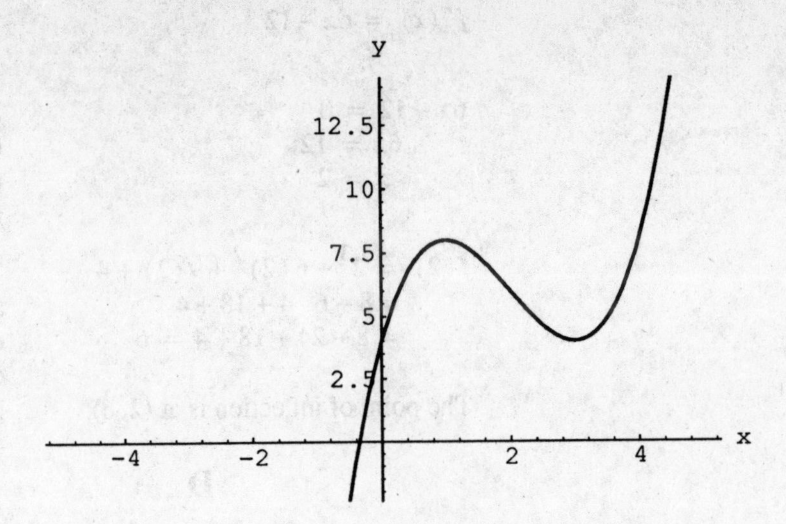

b) $f(x) = \dfrac{2x}{(1-x)^2}$ — Copy the original function.

$f(0) = \dfrac{2(0)}{(1-0)^2} = \dfrac{0}{1} = 0$ — Evaluate the function at zero to find the *y*-intercept.

y-intercept = 0

$$f'(x) = \frac{[2]\,(1-x)^2 - (2x)\,[2\,(1-x)\,(-1)]}{((1-x)^2)^2}$$

Find the first derivative.

$$f'(x) = \frac{2\,(1-x)^2 + 4x\,(1-x)}{(1-x)^4}$$

$$f'(x) = \frac{\cancel{(1-x)}\,[2\,(1-x) + 4x]}{\cancel{(1-x)}\,(1-x)^3}$$

Factor out $(1-x)$.

$$f'(x) = \frac{2-2x+4x}{(1-x)^3}$$

$$f'(x) = \frac{2+2x}{(1-x)^3}$$

$$2+2x = 0$$
$$2x = -2$$
$$x = -1$$

Setting the numerator of the first derivative to zero gives us a zero (critical point).

$$f(-1) = \frac{2(-1)}{(1-(-1))^2} = \frac{-2}{2^2} = -\frac{2}{4}$$

$$= -\frac{1}{2}$$

$\left(-1, -\frac{1}{2}\right)$ is an extreme value.

The *original* function is evaluated at the zero of the first derivative to find the relative extreme value.

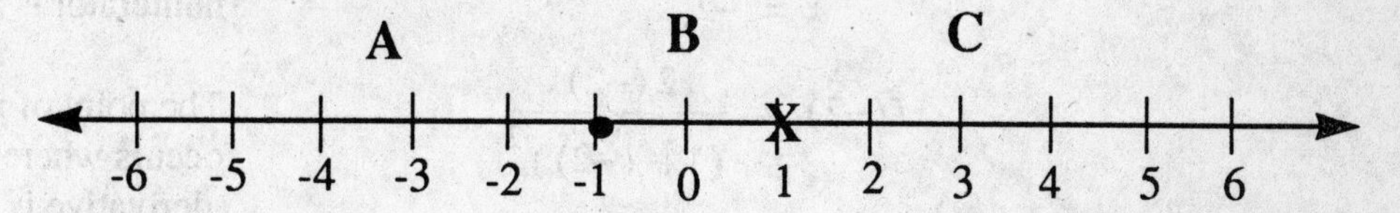

The sign test is used next. The critical points of the first derivative divide the number line into three regions. (The denominator produces a point of discontinuity at $x = 1$.)

region	*A*	*B*	*C*
test point	–2	0	2
sign	–	+	–

The function is decreasing in regions *A* and *C* and increasing in region *B*.

$$f'(-2) = \frac{2+2(-2)}{(1-(-2))^3} = \frac{2-4}{3^3} = \frac{-2}{27} \Rightarrow -$$

$$f'(0) = \frac{2+2(0)}{(1-(0))^3} = \frac{2}{1^3} = 2 \Rightarrow +$$

$$f'(2) = \frac{2+2(2)}{(1-(2))^3} = \frac{2+4}{(-1)^3} = \frac{6}{-1} \Rightarrow -$$

$$f''(x) = \frac{[2](1-x)^3 - [2+2x][3(1-x)^2(-1)]}{((1-x)^3)^2}$$

The second derivative is found.

$$f''(x) = \frac{\cancel{(1-x)^2}[2(1-x)+3(2+2x)]}{\cancel{(1-x)^2}(1-x)^4}$$

$$f''(x) = \frac{2-2x+6+6x}{(1-x)^4} = \frac{8+4x}{(1-x)^4}$$

$$8+4x = 0$$
$$4x = -8$$
$$x = -2$$

The zero of the second derivative occurs when the numerator is zero.

$$f(-2) = \frac{2(-2)}{(1-(-2))^2}$$
$$= \frac{-4}{3^2} = -\frac{4}{9}$$

The point of inflection occurs where the second derivative is zero. The *original* function is evaluated at this point.

$\left(-2, -\frac{4}{9}\right)$ is the point of inflection.

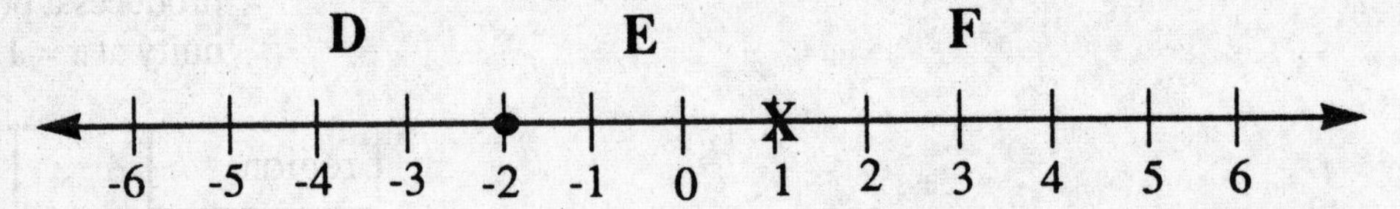

The critical points of the second derivative divide the number line into three regions. (Again, 1 is a point of discontinuity for the second derivative).

region	*D*	*E*	*F*
test point	–3	0	2
sign	–	+	+

$$f''(-3) = \frac{8+4(-3)}{(1-(-3))^4} = \frac{8-12}{(4)^4} = \frac{-4}{256} \Rightarrow -$$

$$f''(0) = \frac{8+4(0)}{(1-0)^4} = \frac{8}{1^4} = 8 \Rightarrow +$$

$$f''(2) = \frac{8+4(2)}{(1-2)^4} = \frac{8+8}{(-1)^4} = \frac{16}{1} \Rightarrow +$$

The signs indicate that the function is concave up in regions *E* and *F* and concave down in region *D*.

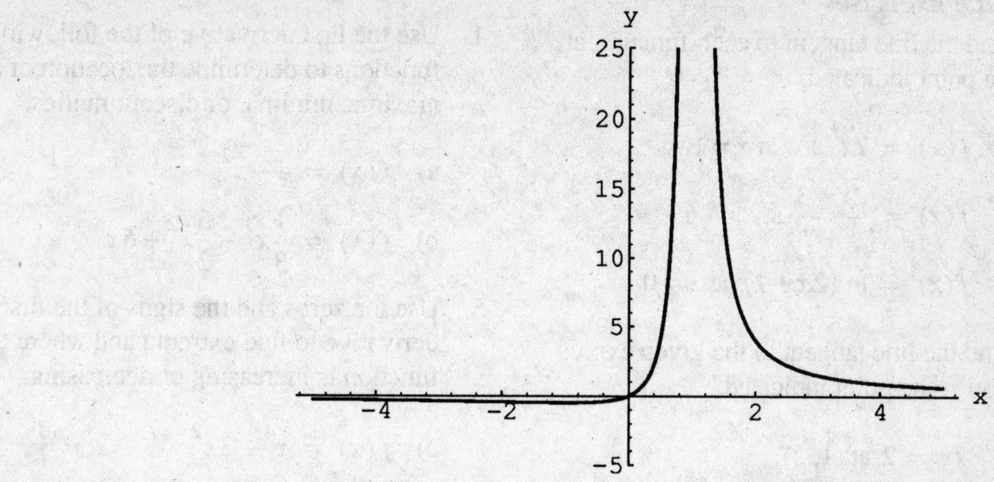

The graph of

$$f(x) = \frac{2x}{(1-x)^2}$$ is above.

Compare the information found during the procedure with the graph.

Practice Exercises

1. Find the line tangent to each function at the point indicated.

   a) $f(x) = 2x^3 + 6$ at $x = 1$

   b) $f(x) = \dfrac{2x-3}{x+7}$ at $x = -6$

   c) $f(x) = \ln(2x+1)$ at $x = 0$

2. Find the line tangent to the given expression at the point indicated.

   a) $xy = 2$ at $(1, 2)$

   b) $y^3 - x = 4$ at $(4, 2)$

   c) $x\ln y = 4$ at $(1, 1)$

3. Determine the critical points of the functions whose first derivatives are given as follows.

   a) $f'(x) = 2x - 7$

   b) $f'(x) = \dfrac{x+1}{4x+5}$

   c) $f'(x) = \dfrac{4x}{\sqrt{3-x}}$

   d) $f'(x) = \dfrac{4}{x^2+1}$

4. Use the first derivative of the following functions to determine the location of any maxima, minima, or discontinuities.

   a) $f(x) = 4 - x^2$

   b) $f(x) = \frac{1}{3}x^3 - \frac{5}{2}x^2 + 6x$

5. Use the zeros and the signs of the first derivative to find extrema and where the function is increasing or decreasing.

   a) $f(x) = x^4 - 2x^2$

   b) $f(x) = \dfrac{3}{x+1}$

6. Find the regions on the graph where the function is concave up or concave down. Determine any points of inflection.

   a) $f(x) = x^3 - 9x^2 + 27x + 3$

   b) $f(x) = \dfrac{6}{x^2}$

Answers

1. a) $6x - y + 2 = 0$

   b) $17x - y + 87 = 0$

   c) $2x - y = 0$

2. a) $2x + y - 4 = 0$

   b) $x - 12y + 20 = 0$

   c) $y - 1 = 0$

3. a) $x = \frac{7}{2}$

   b) $x = -1, x = -\frac{5}{4}$

   c) $x = 0, x = 3$

   d) none

4. a) maximum at $(0, 4)$

   b) maximum at $\left(2, \frac{14}{3}\right)$

   minimum at $\left(3, \frac{9}{2}\right)$

5. a) maximum at $(0, 0)$
   minimum at $(1, -1)$ and at $(1, -1)$

   decreasing $(-\infty, -1)$ $(0, 1)$
   increasing $(-1, 0)$ $(1, \infty)$

   b) no maximum or minimum.

   decreasing $(-\infty, -1)$ $(-1, \infty)$

6. a) concave down $(-\infty, 3)$
   concave up $(3, \infty)$
   point of inflection $(3, 30)$

   b) concave up $(-\infty, 0)$ $(0, \infty)$
   no point of inflection.

5

Applications of Calculus

5.1 MARGINAL COST, MARGINAL REVENUE, AND MARGINAL PROFIT

Introduction

There are many uses of derivatives in practical applications. This chapter will cover some of these uses.

The applications shown are typical of those found in many calculus texts. Although each solution is unique, some general guidelines are offered. The application topics covered are:

a) Marginal cost, revenue, and profit
b) Optimization (maxima and minima)
c) Rates
d) Related rates
e) Elasticity of demand
f) Annuities
g) Compound interest

Marginal Cost, Revenue, and Profit

THE CONCEPT OF MARGINAL VALUES

The amount of cost, revenue, or profit associated with the net output per unit produced or sold is called a *marginal*. The marginal cost of the 31st unit is the total cost of producing 31 units minus the total cost of producing 30 units. Table 5.1 shows the symbols used to represent the typical cost, revenue, and profit equations and the marginal cost, marginal revenue, and the marginal profit equations.

Symbols	Meaning
$C(x)$ or $C(q)$	cost equation
$C'(x)$ or $C'(q)$	marginal cost
$R(x)$ or $R(q)$	revenue equation
$R'(x)$ or $R'(q)$	marginal revenue equation
$P(x)$ or $P(q)$	profit equation
$P'(x)$ or $P'(q)$	marginal profit equation

Table 5.1 Marginal Symbols

Note: Some books use x for their independent variable, while others use q.

EXAMPLE 5.1

a) Find the marginal cost of producing the 20th unit, if the cost equation is $C(x) = 4x + 200$.

b) Find the marginal revenue from the 26th unit, if the revenue equation is $R(x) = 3x^2 - 5x$.

c) Calculate the marginal profit derived from the 8th unit, if the profit equation is $P(q) = q^3 - 2q - 100$.

SOLUTION 5.1

a) $C(x) = 4x + 200$ — Copy the given cost equation.

$$C(20) = 4(20) + 200 = 80 + 200 = \$280$$

$$C(19) = 4(19) + 200 = 76 + 200 = \$276$$

The marginal cost is the *total* cost of producing the 20th unit minus the *total* cost of producing the 19th unit.

$\$280 - \$276 = \$4$ — The marginal cost is \$4.

b) $R(x) = 3x^2 - 5x$ — Copy the given revenue equation.

$$R(26) = 3(26)^2 - 5(26) = 3(676) - 130 = 2028 - 130 = 1898$$
$$R(25) = 3(25)^2 - 5(25) = 3(625) - 125 = 1875 - 125 = 1750$$

The marginal revenue is the *total* revenue derived from the 26th unit minus the *total* revenue derived from the 25th unit.

$\$1898 - \$1750 = \$148$ — The difference is $148. That is the marginal revenue.

c) $P(q) = q^3 - 2q - 100$ — Write down the given profit equation.

$$P(8) = 8^3 - 2(8) - 100 = 512 - 16 - 100 = 512 - 116 = \$396$$
$$P(7) = 7^3 - 2(7) - 100 = 343 - 14 - 100 = 343 - 114 = \$229$$

The marginal profit is the *total* profit earned from the sale of 8 units minus the *total* profit earned from the sale of 7 units.

$\$396 - \$229 = \$167$ — The marginal profit is $167.

THE DERIVATIVE AS THE MARGINAL EQUATION

The marginal can be approximated by the derivative of the original cost, revenue, or profit equation.

The evaluation of the derivative equation for the production unit just prior to the unit of interest gives the marginal value. To find the marginal cost of the 20th unit, we evaluate the derivative at the 19th unit.

Figure 5.1 shows why the derivative gives a good approximation for the difference in total amount for two adjacent units. The total amount of cost, revenue, or profit is read from the y-axis. The number of units is an x-axis value.

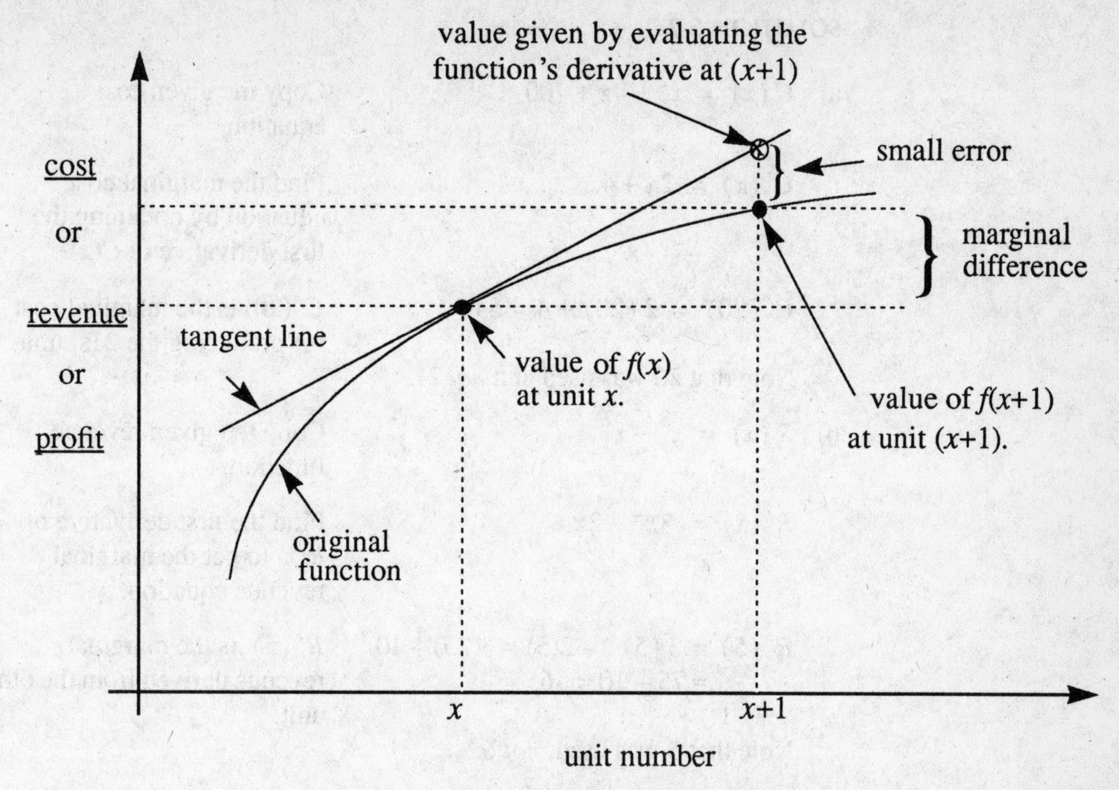

Figure 5.1 Marginal Approximation

The slope from unit 1 to unit 2 for the tangent line in Figure 5.1 is the difference in y-values $f(c)$ and $f(1)$ divided by 1 (2 – 1). $f(c) - f(1)$ is just slightly more than the difference $f(2) - f(1)$. The error is usually small enough to allow the derivative equation to be used. The advantage to using the derivative is that we need to evaluate a function only once, rather than evaluating a function twice *and then* subtracting.

EXAMPLE 5.2

a) Find the marginal cost of the 21st unit for the cost function,

$$C(x) = x^2 + 4x + 700.$$

b) Obtain the marginal revenue derived for the 6th unit if the revenue is

$$R(x) = x^3 - x^2.$$

c) Determine the marginal profit earned on the 36th unit if

$$P(x) = 1000 - x + \frac{x^2}{15}.$$

SOLUTION 5.2

a) $C(x) = x^2 + 4x + 700$ — Copy the given cost equation.

$C'(x) = 2x + 4$ — Find the marginal cost equation by obtaining the first derivative of $C(x)$.

$C'(20) = 2(20) + 4 = \$44$ — $C'(20)$ is the marginal cost for producing the 21st unit.

Note that 20 was used and *not* 21.

b) $R(x) = x^3 - x^2$ — Copy the given revenue function.

$R'(x) = 3x^2 - 2x$ — Find the first derivative of $R(x)$ to get the marginal revenue equation.

$R'(5) = 3(5)^2 - 2(5) = 3(25) - 10$
$= 75 - 10 = \$65$ — $R'(5)$ is the marginal revenue derived from the 6th unit.

Note that 5 was used, *not* 6.

c) $P(x) = 1000 - x + \dfrac{x^2}{15}$ — Copy the given profit function.

$P'(x) = -1 + \dfrac{2x}{15}$ — Find the first derivative of $P(x)$ to get the marginal profit equation.

$P'(35) = -1 + \dfrac{2(35)}{15}$ — The profit earned on the 36th unit is the marginal profit at the 35th unit.

$= -1 + 2\left(\dfrac{7}{3}\right) = -1 + \dfrac{14}{3}$

$= -\dfrac{3}{3} + \dfrac{14}{3} = \dfrac{11}{3} \cong \$3.66$

Note that 35 was used instead of 36.

DEMAND AND REVENUE

The demand for a certain product is regulated by the price charged for that product. Generally, as the price for a certain product increases, the demand decreases. The supply will increase if production levels remain the same.

The *demand function* or *demand equation* is written as $p = f(x)$, where p is the price of the item and x is the units sold or "demanded." When the demand equation is written in the form of the *price* of each unit produced, then the *revenue* can be found by using the demand equation.

Total revenue can be determined by knowing the demand equation and the total units sold. The equation for revenue, $R(x)$, is *total units sold* times *price of each unit.*

The symbolic equation for revenue, then, is:

$$R(x) = [x] \cdot [p(x)]$$

where x is units sold and $p(x)$ is the demand function.

We can use the above information in problems involving demand and revenue.

Some texts set the price equal to some form of the units sold, such as $p = x^2 - 2x$ or $p = \frac{x}{\sqrt{x+2}}$. Other authors use the demand in terms of price, giving equations like $x = \frac{p}{p^2 - 1}$ and $x = \sqrt{p+50}$.

EXAMPLE 5.3

a) Suppose the demand function is given by $p = \frac{\ln x}{x}$. Determine the rate of change in price per unit sold.

b) The demand equation for electric fans is: $x = 2000 - 5p^2$. Find the rate of change of units sold per unit price change.

c) The demand equation for weekly sales of a product is $x = 150 - \frac{300}{p+2}$. Find the marginal revenue equation.

d) The annual sales for a small computer store is $p(x) = 1500 - x$ where x is the number of units sold and p is the price per unit. Find the marginal revenue at 100 units sold.

SOLUTION 5.3

a) $p = \frac{\ln x}{x}$ — The given function is the price of each unit in terms of units sold (x).

$$p' = \frac{\left[\frac{1}{x}\right] x - \ln x\,[1]}{x^2}$$

The first derivative gives us a rate.

$$p' = \frac{\frac{x}{x} - \ln x}{x^2} = \frac{1 - \ln x}{x^2}$$

The rate of change of price per unit sold is given by the first derivative function. dp/dx is Leibniz notation for instantaneous change in price per unit sold.

$$p' = \frac{dp}{dx} = \frac{1 - \ln x}{x^2}$$

Note that an increase in units sold will cause $p'\left(\frac{dp}{dx}\right)$ to have a negative change. That is, the price will drop.

b) $x = 2500 - 5p^2$

The demand equation gives units sold, x, in terms of price, p.

$$x' = \frac{dx}{dp} = 0 - 10p$$

The first derivative of the demand equation shows the change of the number of units sold per unit change in price dx/dp.

$$\frac{dx}{dp} = -10p$$

The result shows that for every unit price increase, 10 less units are sold.

Note: The answer to Example 5.3(a) used dp/dx as the first derivative. The answer to Example 5.3(b) used dx/dp. The symbols look similar but have very different meanings.

c) $x = 150 - \frac{300}{p+2}$

The demand equation is shown as the units sold, x, in terms of price, p.

$$x = 150 - \frac{300}{p+2}$$

Remembering that revenue, $R(x)$, is in terms of x, we solve for p in terms of x.

$$x - 150 = -\frac{300}{p+2}$$

Subtract 150 from both sides.

$$-\frac{x - 150}{300} = \frac{1}{p+2}$$

Divide both sides by –300.

$$\frac{-300}{x - 150} = \frac{p+2}{1}$$

Invert both sides of the equal sign.

$$\frac{-300}{x-150} - 2 = p$$ Subtract 2 from both sides.

$$R(x) = x \cdot p$$ The equation for $R(x)$ is $x \cdot p$.

$$R(x) = x \cdot \left(\frac{-300}{x-150} - 2\right)$$

$$R(x) = \frac{-300x}{x-150} - 2x$$ $R(x) = x \cdot p$.

$$R'(x) = \frac{[-300]\,(x-150) - (-300x)\,[1]}{(x-150)^2} - 2$$

The marginal revenue is the first derivative of the revenue equation.

$$R'(x) = \frac{-300x + 45{,}000 + 300x}{(x-150)^2} - 2$$

$$R'(x) = \frac{45{,}000}{(x-150)^2}$$

d) $p(x) = 1500 - x$ The demand equation (written as p in terms of x) is given.

$$R(x) = x \cdot p$$ $R(x) = x \cdot p$.

$$= x(1500 - x)$$

$$= 1500x - x^2$$

$$R'(x) = 1500 - 2x$$ The marginal revenue is the first derivative of the revenue function ($R(x)$).

$$R'(100) = 1500 - 2(100)$$ The marginal revenue at 100 units sold is $R'(100)$.

$$R'(100) = 1500 - 200$$

$$R'(100) = \$1300$$

The marginal revenue at 100 units sold is $1300.

PROFIT EQUATIONS AND THE BREAK-EVEN POINT

The profit for the sale of an item or a service is the total revenue obtained minus the cost of production. The formula for profit is:

5.1 Profit Equation

$$P(x) = R(x) - C(x)$$

where $P(x)$ is the profit for x units, $R(x)$ is the revenue for x units and $C(x)$ is the cost for x units.

When the incoming revenue is equal to the cost of production, the profit is zero. The zero point for profit is known as the **break-even point**. The formula for the break-even point is as follows:

5.2 Break-Even Point Equation

a) $P(x) = R(x) - C(x) = 0$

b) $R(x) = C(x)$

where $P(x)$ is profit, $R(x)$ is revenue, and $C(x)$ is cost.

EXAMPLE 5.4

a) The demand equation for the production of metal awnings is $p = 1000 - 2x$ and the cost of production is $C(x) = 20{,}000 + 100x$. Find the marginal profit for the 51st unit.

b) The profit earned for each visit by a customer to a photography studio is found according to the equation $P(x) = 600 + 10x - x^2$, where x is the number of customers per week. Find the number of customers needed to break even for the week.

c) The demand function for the weekly sales of x books is $p = 400 - x$. The cost function is $C(x) = 1500 + 2x$. Find the break-even point for sales and the marginal profit for 101 books.

SOLUTION 5.4

a) $C(x) = 20{,}000 + 100x$ — Copy the given equations.

$p = 1000 - 2x$

$P(x) = R(x) - C(x)$ — The profit equation is the difference between the revenue and the cost.

we need $R(x)$ and $C(x)$

$R(x) = xp = x(1000 - 2x)$ — The revenue is x times the demand equation.

$= 1000x - 2x^2$

$P(x) = R(x) - C(x)$ — The profit equation can be found.

$P(x) = (1000x - x^2) - (20{,}000 + 100x)$

$P(x) = 1000x - x^2 - 20{,}000 - 100x$

$P(x) = 1000x - 100x - x^2 - 20{,}000$

$P(x) = 900x - x^2 - 20{,}000$

$P'(x) = 900 - 2x - 0$

$P'(x) = 900 - 2x$ — The marginal profit equation is the first derivative of the profit equation.

$P'(50) = 900 - 2(50)$ — We use $x = 50$ when we want to find the marginal point for the 51st unit.

$P'(50) = 900 - 100$

$P'(50) = \$800$

The marginal profit for the 51st unit produced and sold is $800.00.

b) $P(x) = 600 + 10x - x^2$ — The profit equation based upon the number of customers per week, x, is copied.

$P(x) = 600 + 10x - x^2 = 0$ — The break-even point is when the profit function equals zero.

$600 + 10x - x^2 = 0$ — One way to solve this quadratic equation is to determine the factors and set them equal to zero.

$(30 - x)(20 + x) = 0$

$30 - x = 0 \qquad 20 + x = 0$

$-x = -30 \qquad x = -20$

$x = 30$

$x = 30 \qquad x = -20$ — The two solutions are given. Since -20 is not in the domain for x (the numbers of customers is 0 or more) the correct answer is $x = 30$.

Thirty (30) customers are needed before the photography studio can break even.

c) $C(x) = 1500 + 2x$

$p = 400 - x$

The cost function, $C(x)$ and the demand function, p, are copied.

$P(x) = R(x) - C(x)$

The profit must be determined from the difference of the revenue function, $R(x)$, and the cost function, $C(x)$.

$R(x) = xp$

$R(x) = x(400 - x)$

$R(x) = 400x - x^2$

First we must determine $R(x)$ from the demand function.

$P(x) = R(x) - C(x)$

The profit function, $P(x)$, can now be determined.

$P(x) = (400x - x^2) - (1500 + 2x)$

$P(x) = 400x - x^2 - 1500 - 2x$

$P(x) = 400x - 2x - x^2 - 1500$

$P(x) = 398x - x^2 - 1500$

break-even

$P(x) = 398x - x^2 - 1500 = 0$

Break-even occurs at $P(x) = 0$. We are unable to factor this trinomial through the usual means.

$$x = \frac{-b \pm \sqrt{b^2 - 4ac}}{2a}$$

The quadratic formula must be used.

$$P(x) = \underset{a}{-1x^2} + \underset{b}{398x} - \underset{c}{1500}$$

$a = -1, b = 398, c = -1500$

The values of a, b, and c are found from the trinomial numerical coefficients.

$$x = \frac{-(398) \pm \sqrt{(398)^2 - 4(-1)(-1500)}}{2(-1)}$$

$$x = \frac{-(398) \pm \sqrt{158,404 - 6000}}{-2}$$

The substituted quadratic formula is found.

$$x = \frac{-398 \pm \sqrt{152,404}}{-2}$$

$$x = \frac{-398 \pm 390.4}{-2}$$

There are two choices after this point.

$$x_1 = \frac{-398 + 390.4}{-2} \qquad x_2 = \frac{-398 - 390.4}{-2}$$

$$x_1 = \frac{-7.6}{-2} \qquad x_2 = \frac{-788.4}{-2}$$

$$x_1 = 3.8 \approx 4 \qquad x_2 = 394.2 \approx 394$$

The solutions of $x = 4$ books and $x = 394$ books are the two break-even points.

$$P(x) = 398x - x^2 - 1500$$
$$P'(x) = 398 - 2x + 0$$
$$P'(x) = 398 - 2x$$

The marginal profit is found from the first derivative of the profit function.

$$P'(100) = 398 - 2(100)$$
$$P'(100) = 398 - 200$$
$$P'(100) = \$198$$

The marginal profit for the 101st unit is the marginal profit at the 100th unit.

The marginal profit at 101 books is $198.00.

5.2 OPTIMIZATION

In chapter 4 we used the first derivative to find local maximum and local minimum values. Finding the most extreme values in a local interval is the process used in "optimization." There are generally two types of optimization problems shown in business calculus texts. These types are:

a) Geometric problems
b) Business/Life Science applications

Geometric Problems

The geometric problems usually involve maximizing an area or volume while using a given amount of material at hand. The use of geometric shapes usually requires that a sketch be drawn to display the given information. Also, we must remember formulas for the area of a rectangle and circle and the formulas for the volume of a box or cylinder. These formulas are given below:

Use	Formula
Circumference of a circle	$C = 2\pi r$

Use	Formula
Perimeter of a rectangle	$P = 2l + 2w$
Area of a rectangle	$A = l \times w$
Area of a circle	$A = \pi r^2$
Volume of a box	$V = l \times w \times h$
Volume of a cylinder	$V = \pi r^2 \times h$

where A = area, V = volume, l = length, w = width, h = height, r = radius, $\pi \approx 3.14$, P = perimeter, and C = circumference.

An important point to remember is that we are looking for *local* maximum or *local* minimum values. This means that we must restrict the interval in which we will look for solutions to the problem. Some problems may have two or more *mathematical* solutions, from which we choose the solution that makes the most sense. For example, a problem might yield two solutions. One solution might represent a positive length, while the other represents a negative length. Obviously, a negative length does not exist in a real sense and this solution must be discarded.

The guidelines for solving geometrical optimization problems are as follows:

(A) After reading the problem, draw a sketch that labels everything that is known.

(B) In conjunction with the labels from step A, assign symbols to the significant variables in the problem.

(C) Determine the relations between the variables and try to establish an equation that defines the variable to be optimized in terms of *one* other variable.

(D) Determine the interval of importance by observing the physical and the numerical *constraints* of the problem.

(E) Find the first derivative of the equation formed in step C and solve at zero. Test answers against the interval determined in step D.

EXAMPLE 5.5

a) A city-dweller has purchased a puppy and 400 feet of fencing. She intends to build a rectangular pen for the puppy. However, she wants the area enclosed by the fence to be as large as possible so that the puppy has room to play. What should the length and width of the pen be?

b) A manufacturer makes steel enclosures for electrical connections. The enclosures, excluding a face plate, are to be made from a piece of metal 16 inches square. The open top is to be made by cutting a square from each corner and bending up the resulting tabs. What size square should be cut from the corner to make the volume of the electrical enclosure maximum?

c) A company that makes road flares wishes to package these flares in a PVC cylinder with plastic endcaps. Each cap costs as much per square inch as the tube material. The volume must be 128π cubic inches. Find the dimensions of the cylinder that will be the most economical. The height of the cylinder must be at least 6 inches to enclose the flares.

SOLUTION 5.5

a)

dog pen (rectangle with sides w and l)	(A) Sketch the geometric figure and label.
Area of dog pen, $A = l \cdot w$ Perimeter of dog pen, $P = 2l + 2w$	(B) Assign variables.
length of fence = 400 feet = perimeter.	(C) The area, A, is to be optimized, so we must define the area in terms of *one* other variable.
$400 = 2l + 2w$	Set the length of fence equal to perimeter.
$200 = l + w$	Divide by 2.
$l = 200 - w$	Solve for l.
$A = l \cdot w$ $A = (200 - w) \cdot w$	Substitute for l in area formula.
$A = 200w - w^2$	

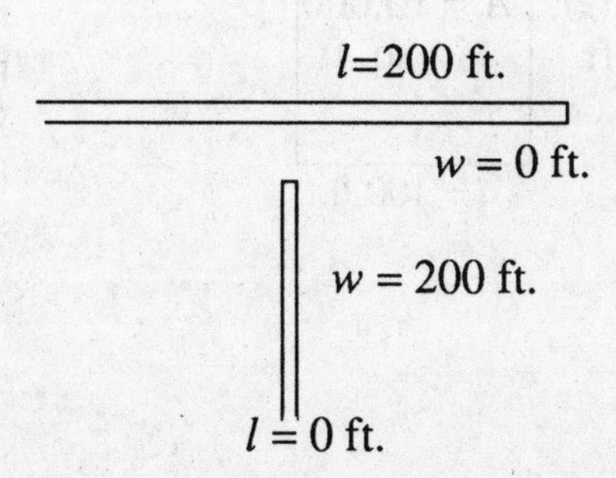

(D) Determine the physical limits of the solution interval.

$0 \le w \le 200$	The width can be no narrower than 0 feet and no wider than 200 feet. (This occurs when the fencing is merely folded in half.)
$\frac{dA}{dw} = 200 - 2w$	(E) Find $\frac{dA}{dw}$ and set to 0.
$0 = 200 - 2w$ $2w = 200$ $w = 100$ feet	Solve for w. $w = 100$ is where the maximum area occurs. If $w = 0$ or 200, the area enclosed is zero. All other values of w in this interval give a smaller enclosed area. Check the given solution *and* the endpoints of the interval to see which works best. Other points in the interval need *not* be checked.
$200 - w = l$ $200 - 100 = l$ $100 = l$	The length, l, is determined from the perimeter equation in part B.
$A = l \cdot w$ $A = (100) \cdot (100)$ $A = 10{,}000$ square feet	The maximum area is found by multiplying the length times the width.
$w = 100$ ft. | $A = 10{,}000$ sq. ft. | $l = 100$ ft.	The member of the rectangle family that encloses the most area for a given perimeter is a *square*. You will find that a square is always the maximum solution.

b)

(A) Sketch the geometric figure and label.

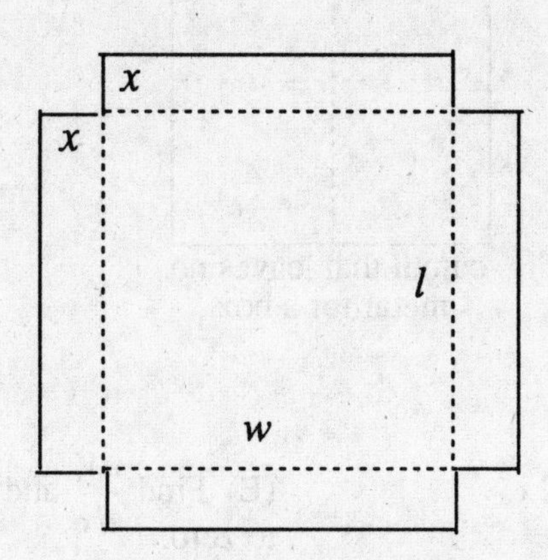

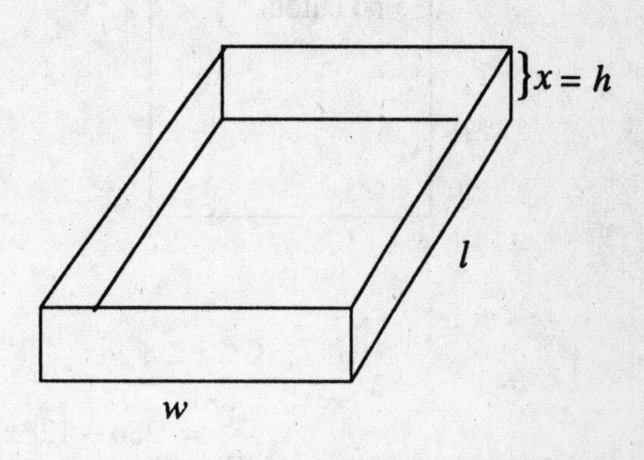

Volume of open box = $l \cdot w \cdot h$

$l = 16 - 2x = w$

$h = x$

$V = l \cdot w \cdot h$

(B) Assign variables. The length, l, and the width, w are equal to 16 minus two times x (the length of the side of the square cutout).

The height will equal x.

(C) The volume, V, is to be optimized, so we must find the volume in terms of *one* variable.

$l = 16 - 2x$
$w = 16 - 2x$
$h = x$
$V = (16 - 2x)\,(16 - 2x)\,(x)$

The length, width, and height are all in terms of x.

Rewrite the volume in terms of x.

$V = (256 - 64x + 4x^2)\,(x)$

$V = 256x - 64x^2 + 4x^3$

(D) Determine the physical limits of the solution interval for x. x can be no less than zero (no cutout) and no more than one-half of the width of the metal sheet (or 8 inches).

no cutout

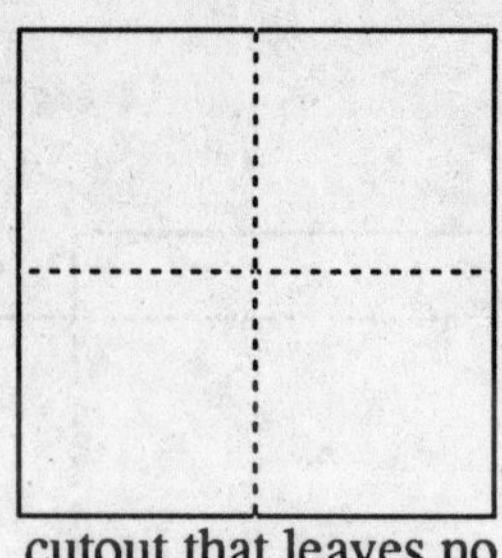

cutout that leaves no metal for a box

$$0 \le x \le 8$$

$$\frac{dV}{dx} = 256 - 128x + 12x^2$$

(E) Find $\frac{dV}{dx}$ and set equal to zero.

$$0 = 12x^2 - 128x + 256$$

Divide by 4.

$$0 = 3x^2 - 32x + 64$$

$$0 = (3x - 8)(x - 8)$$

Factor trinomial.

$$0 = 3x - 8 \qquad x - 8 = 0$$
$$3x = 8 \qquad x = 8$$
$$x = \frac{8}{3}$$

Solve for solution.

$$x = \frac{8}{3}$$

The value of $x = 8$ cannot be used because that leaves no metal to make a box.

$$h = x = \frac{8}{3} \text{ inches}$$

$x = 8/3$ is the optimum solution.

$$l = 16 - 2x = 16 - 2\left(\frac{8}{3}\right) = 10\frac{2}{3} \text{ inches}$$

$$w = l = 10\frac{2}{3} \text{ inches}$$

The length, height, and width can be found from relating each to x.

c)

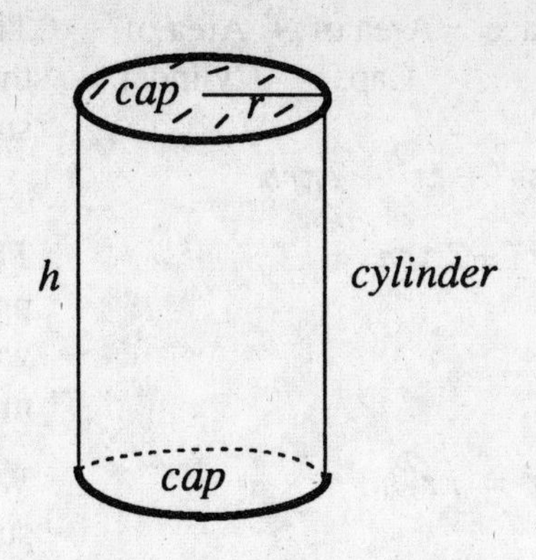

(A) Sketch the geometric figure and label.

h = height of cylinder
r = radius of cap

Volume = $V = \pi r^2 \cdot h$

area of circle — height of cylinder

(B) Assign symbols for the significant variables in the problem.

"unrolled" cylinder

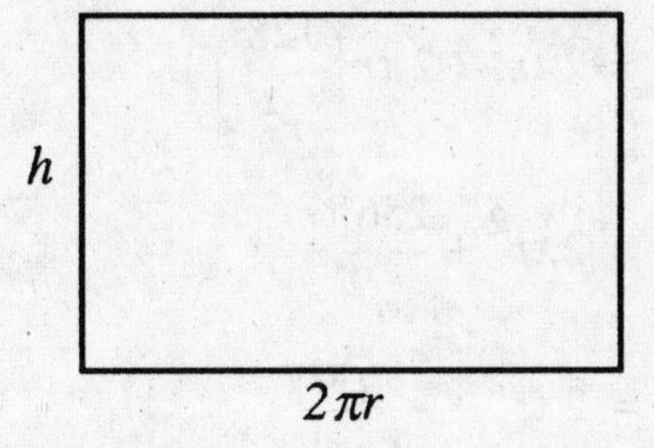

Surface Area = SA

$SA_{cylinder} = 2\pi r \cdot h$

circumference of circle $2\pi r$ — height h

(C) Since cost is involved in this problem, we must determine the relationship to the cost per unit surface area of the container. The surface area of the cylinder (found by cutting the cylinder lengthwise and flattening it) is the cylinder circumference times cylinder height.

Area of endcap = πr^2

The area of the two endcaps is just the area of their circular surface.

Total Surface Area = Area of Caps + Area of Cylinder	The total surface area, *SA*, is the area of each cap plus the surface area of the cylinder.
$SA_{\text{total}} = \pi r^2 + \pi r^2 + 2\pi rh$	
$SA = 2\pi r^2 + 2\pi rh$	The area is *directly* proportional to the cost, so we can minimize cost by minimizing surface area, *SA*.
$V = 128\pi = \pi r^2 h$	We would like to get the surface area, *SA*, in terms of *one* variable. We can do that if we bring the volume equation into play.
$\frac{128\pi}{\pi r^2} = h$	Solve for *h* in terms of *r*. The result will be substituted for *h* in the surface area equation.
$SA = 2\pi r^2 + 2\pi r\left(\frac{128}{r^2}\right)$	The surface area equation can now be written in terms of radius *only*.
$SA = 2\pi r^2 + \frac{256\pi r}{r^2}$	
$SA = 2\pi r^2 + \frac{256\pi r}{r \cdot r}$	
$SA = 2\pi r^2 + \frac{256\pi}{r}$	
h can be no less than 6.	(D) The interval begins at zero. There can be no radius less than zero. Since the height must be at least 6 inches, the radius can be no more than that related by the equation for volume at 128π cubic inches.
$V = \pi r^2 h$	
$128\pi = \pi r^2 h$	
$128\pi = \pi r^2 (6)$	
$\frac{128\pi}{6\pi} = r^2$	
$\frac{128}{6} = \frac{64}{3} = r^2$	

$$\frac{8}{\sqrt{3}} = \frac{8\sqrt{3}}{3} = r$$

$$0 \le r \le \frac{8\sqrt{3}}{3}$$

The interval must be between 0 and $(8\sqrt{3})/3$ (approximately 4.62).

$$SA = 2\pi r^2 + \frac{256\pi}{r}$$

(E) Find the first derivative of the surface area with respect to r and set it equal to zero.

$$\frac{dSA}{dr} = 4\pi r - \frac{1(256)\pi}{r^2}$$

$$0 = 4\pi r - \frac{256\pi}{r^2}$$

$$\frac{0}{4\pi} = \frac{4\pi r}{4\pi} - \frac{256\pi}{r^2} \cdot \frac{1}{4\pi}$$

Divide by 4π.

$$0 = r - \frac{64}{r^2}$$

Add $64/r^2$ to both sides of the equation.

$$\frac{64}{r^2} = r$$

Multiply by r^2.

$$64 = r^3$$

Find the cube root of both sides.

$$4 = r$$

The answer, $r = 4$, falls in the interval for

$$0 \le r \le \frac{8\sqrt{3}}{3} \approx 4.62.$$

$$V = 128\pi = \pi(r^2)h$$

$$128\pi = \pi(4^2)h$$

$$128\pi = 16\pi h$$

$$\frac{128\pi}{16\pi} = h$$

$$8 = h$$

The height of 8 inches and a radius of 4 inches gives us the most economical cylinder.

Business Applications

This class of problems is, perhaps, easier to complete since we are given the formulas to be used. The main task is to find the first derivative of the given formula and solve it at zero. The result is the solution to the problem. The procedure for solving those problems with given formulas consists of three steps.

(A) Find the first derivative of the given formula.
(B) Solve the first derivative at zero.
(C) Compare the solution against the endpoints of the interval (usually given).

EXAMPLE 5.6

a) The profit function for the production of automobile radiators is $P(x) = 500x - x^2$ where x is the number of radiators produced each week $(0 \leq x \leq 500)$. How many radiators should be produced to yield the maximum profit?

b) A diesel truck burns fuel at a rate of

$$g(x) = \frac{2900 + x^2}{x}$$

gallons per mile when traveling at x miles per hour. Since the minimal cost is directly related to the minimization of fuel use, find the speed at which fuel use will be minimized.

c) The concentration of a certain drug in the bloodstream at time, t, after ingestion is given by

$$h(t) = 6t^{2/3} - 4t \qquad 0 \leq t \leq 3$$

where t is in hours. How many hours after ingestion is the concentration in the bloodstream maximum?

d) Currently, the owner of a small business sells 60 wooden cabinets per week at $40 per cabinet. The owner has calculated that, if the price of each cabinet is raised one dollar, the number of cabinets sold drops by one. What price should be charged to produce the maximum revenue for the week.

SOLUTION 5.6

a) $P(x) = 500x - x^2$ — Copy the given profit function.

$P'(x) = 500 - 2x$ — (A) Find the first derivative of the profit function, $P(x)$.

$$P'(x) = 0 = 500 - 2x$$
$$500 = 2x$$

(B) Set $P'(x) = 0$ and solve.

$$250 = x$$

The extreme value is 250.

$$P(0) = 500(0) - 0^2 = 0$$
$$P(250) = 500(250) - (250)^2 = 62{,}500$$
$$P(500) = 500(500) - 500^2 = 0$$

(C) The endpoints are 0 and 500. The value at $P(250)$ is the largest. Therefore, the value of 250 produces the maximum profit.

b) $$g(x) = \frac{2900 + x^2}{x}$$

Copy the given equation.

$$g(x) = 2900x^{-1} + x$$

$$g'(x) = (-1)(2900)x^{-2} + 1$$
$$g'(x) = \frac{-2900}{x^2} + 1$$

(A) Find the first derivative of the given equation.

$$\frac{-2900}{x^2} + 1 = 0$$

(B) Set the first derivative equal to zero. Solve for x.

$$\frac{-2900}{x^2} = -1$$

$$-2900 = -x^2$$

$$2900 = x^2$$

$$x = +53.85 \text{ or } -53.85$$

$$\lim_{x \to 0} g(x) = \lim_{x \to 0} \frac{2900 + x^2}{x} = \infty$$

$$\lim_{x \to \infty} g(x) = \lim_{x \to \infty} \frac{2900 + x^2}{x} = \infty$$

$$g(53.85) = \frac{2900 + (53.85)^2}{53.85}$$

$$g(53.85) = 107.7$$
minimum value

(C) The endpoints have not been given, but the truck must have a speed greater than 0. The upper limit can approach infinity. Since both endpoints cannot be reached, we use limits. The value at 53.85 is the *minimum*, since the endpoints tend toward infinity.

c) $h(t) = 6t^{2/3} - 4t$ — Copy the given function.

$h'(t) = \frac{2}{3}(6)t^{-1/3} - 4$ — (A) Find the first derivative of the given function.

$h'(t) = 4t^{-1/3} - 4$

$h'(t) = \frac{4}{t^{1/3}} - 4$

$\frac{4}{t^{1/3}} - 4 = 0$ — (B) Solve the first derivative at 0.

$\frac{4}{t^{1/3}} = 4$

$4t^{-1/3} = 4$

$t^{-1/3} = 1$

$t = 1$

$h(0) = 6 \cdot 0^{2/3} - 4(0)$
$= 0 - 0 = 0$ — (C) The given endpoints are 0 and 3. The endpoints are compared to the value of $t = 1$.

$h(1) = 6 \cdot 1^{2/3} - 4(1) = 6 - 4 = 2$

$h(3) = 6 \cdot 3^{2/3} - 4(3) = 12.52 - 12$
$= 0.52$

d) current revenue is:
cabinets sold × price
$R = 60 \times \$40 = \$2400$

This problem does not have a given equation, but one can be found by comparing the current revenue against the change in revenue if the price is raised.

$R(x) = (60 - x)(40 + x)$
(60 − x): lose sale of one cabinet; (40 + x): add one dollar to price

The revenue function is found by multiplying the two binomials.

$R(x) = 2400 + 20x - x^2$

$R'(x) = 20 - 2x$ — (A) Find the first derivative of the revenue function.

$20 - 2x = 0$ $20 = 2x$ $x = 10$	(B) Solve the first derivative at zero.
$R(0) = (60)(40) = \$2400$	(C) The limits for this function are $x = 0$, which represent no change and $x = 60$, which means no cabinets are sold.

$$R(10) = (60 - 10)(40 + 10) = (50)(50) = \$2500$$

$$R(60) = (60 - 60)(40 + 60) = (0)(100) = \$0$$

A price of \$50 will produce maximum revenue.

5.3 RATES

Many algebra students are aware of the distance formula:

$$D = R \cdot T$$

where D is distance, R is rate, and T is time.

If this equation is solved for R, we note the rate is expressed as D/T, or distance divided by time. All rates are expressed per unit time, with time in the divisor.

In calculus, rates are instantaneous so we express them in Leibniz notation as $d\,[\text{any variable}]\,/dt$. The variable is anything that changes with time. Therefore, when we see an expression such as dS/dt or dx/dt, we know that the expression represents a rate. Once an equation is differentiated with respect to time, a rate has been established. If the values for certain variables in the equation are known, we can determine the numerical value of the instanteous rate.

EXAMPLE 5.7

a) The height (in meters) of a particle moving in a gravity field is given by the equation

$$h(t) = 32t - t^2 \qquad t \text{ is in seconds.}$$

Find the rate of change in height at the instant when $t = 2.7$ seconds.

b) The price of a stock varies according to the equation

$$p(t) = 10 + 2\sqrt{t}$$

where: p is in dollars per share
t is time in months from the stock's first offering.

At what rate is the price increasing at the 4th month after the stock is first offered?

c) City planners have determined that a city is growing as given by the equation

$$P(t) = 10{,}000\,(e^{0.2t})$$

where P is the population and t is the time in years since 1980. Find the rate of increase in population in 1985.

d) The concentration of a certain drug in the bloodstream t minutes after it is injected is:

$$C(t) = 10(\ln t) - t^2$$

where C is the concentration in micrograms/deciliters. Find the change in concentration three minutes after injection.

SOLUTION 5.7

a) $h(t) = 32t - t^2$

$$\frac{dh}{dt} = 32 - 2t = h'(t)$$

The rate of change in height is the first derivative of the height function.

$$h'(2.7) = 32 - 2(2.7) = 32 - 5.4 = 26.6 \text{ m/sec}$$

The rate of change of height for $t = 2.7$ is the first derivative evaluated at $t = 2.7$ seconds.

b) $p(t) = 10 + 2\sqrt{t} = 10 + 2t^{1/2}$

$$\frac{dp}{dt} = p'(t) = 2\left(\frac{1}{2}\right)t^{-1/2} = t^{-1/2}$$

The rate of change in price is the first derivative of the price equation.

$$p'(t) = \frac{1}{\sqrt{4}} = \frac{1}{2} \text{ dollars/month}$$

The rate of change in price at the 4th month after first issue is the first derivative evaluated at $t = 4$.

c) $P(t) = 10{,}000\,(e^{0.2t})$

$$P'(t) = 10{,}000\,(e^{0.2t})\,(0.2) = 2000e^{0.2t}$$

The change in population is the first derivative of the population equation.

$$P'(5) = 2000e^{0.2(5)}$$
$$= 2000e^{1}$$
$$\cong 2000\,(2.718)$$
$$\cong 5436 \text{ people/year}$$

The change in population per year at the 5th year after 1980 (1985) is the derivative evaluated at $t = 5$.

d) $C(t) = 10\ln t - t^2$

The change in concentration is the first derivative of the concentration equation.

$$C'(t) = 10\left(\frac{1}{t}\right) - 2t$$
$$= \frac{10}{t} - 2t$$

$$C'(3) = \frac{10}{3} - 2(3)$$
$$= 3.33 - 6$$
$$= -2.67\ \frac{\text{micrograms}}{\text{deciliter}}$$

The change in concentration 3 minutes after injection is the first derivative evaluated at $t = 3$.

The negative sign means that the concentration is dropping.

5.4 RELATED RATES

As the previous section demonstrated, rates are a variation of some variable with respect to the passing of time. All of the formulas in Example 5.7 were expressed as a function of time. We can evaluate a formula not normally expressed in temporal terms by having a time element introduced by implicitly differentiating the formula with respect to time.

The result is a formula that contains two or more rates. These formulas that "relate" rates through an equal sign are known as *related rates* formulas.

The solution of a problem that involves related rates normally can be found using the following sequence.

(A) If the problem details physical interactions, draw a sketch that shows how items interact. Decide what letters are to be used for variables. Label the sketch.

(B) Assign as many given numerical values to variables as you can. All rates are labeled.

$$\frac{d\,[\text{some variable}]}{dr}$$

The unknown will typically be a rate. Note which rate you are looking for.

(C) Choose an equation that relates any variables that are *not* rates. This is the *static* equation.

(D) Implicitly differentiate the static equation with respect to time. You are setting the static equation in "motion." The "moving" equation is a *dynamic* equation and it relates the rates needed for this problem.

(E) Take a "snapshot" of the dynamic equation by evaluating it at the appropriate variable and rate values found in step (B). **Important**: You may find that you need more variable values at this step. You must go back to steps (A) and (B) to see if any geometric relationships can give you other variable values.

(F) Complete the problem by evaluating all known quantities to obtain the desired solution.

EXAMPLE 5.8

a) An unfortunate accident has caused a rupture in the hull of a petroleum tanker. The resulting oil slick is spreading in a circular manner as it radiates from the oil taken at the center of the circle. If the radius of this circle is increasing at 0.2 kilometers per hour, how fast is the area covered by the oil slick expanding when the radius is 1 kilometer?

b) A spherically shaped star is uniformly expanding at the rate such that its radius is increasing at 100 kilometers per second. How fast is the volume of the star increasing when its diameter is 400,000 kilometers? For a sphere $V = \frac{4}{3}\pi r^3$, where r is the radius and V is the volume.

c) A youthful observer watches his toy rocket rise at the rate of 20 feet/second. The youth stands 300 feet from the rocket launchpad as he watches the rocket rise. How fast is the line-of-sight distance increasing between him and the rocket when the rocket is 400 feet in the air? Assume the flight path is perfectly vertical.

d) A grass fire is covering a rectangularly shaped area. It is advancing along two fronts, a northern front and an eastern front. If the northern front is moving at 0.3 kilometers per hour and the eastern front is moving at 0.2 kilometers per hour, how much is the area increasing when the eastern boundary is 0.5 kilometers in length and the northern boundary is 0.8 kilometers in length?

SOLUTION 5.8

a)

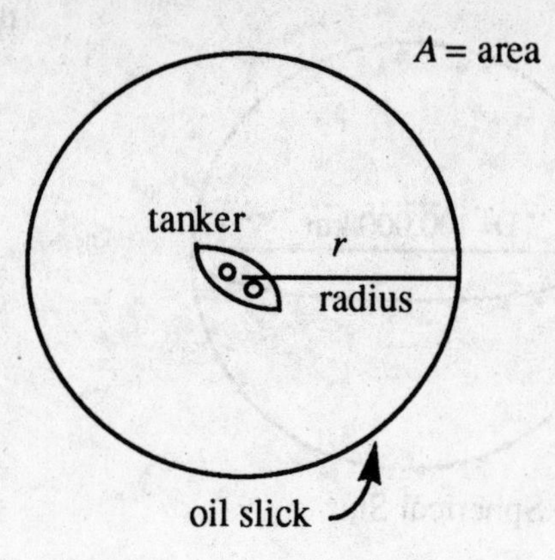

(A) Draw a sketch and select alphabetical labels.

A = area

$\frac{dr}{dt}$ = change in radius = 0.2 km/hr

$\frac{dA}{dt}$ = change in area = ?

r = radius at desired moment = 1 km

(B) The sketch is labeled. The known values are listed.

The static equation is the area of a circle. $A = \pi r^2$

(C) Choose the "static" equation.

$$\frac{dA}{dt} = 2\pi r \frac{dr}{dt}$$

(D) Form the "dynamic" equation by implicit differentiation with respect to time.

$$\frac{dA}{dt} = 2\pi\,(1 \text{ km})\left(0.2 \ \frac{\text{km}}{\text{hr}}\right)$$

(E) Take a "snapshot" at the instant when the radius is 1 km.

$$\frac{dA}{dt} = 2\pi\,(1)\ (0.2) = 0.4\pi = 1.257$$

(F) All values needed are known. Multiply to find the answer.

The change in area, at the instant the radius is 1 kilometer, is 1.257 sq. km/hr.

b)

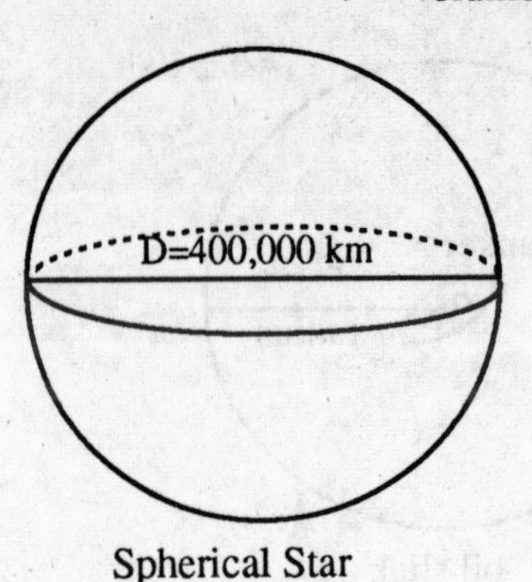

Spherical Star

(A) Draw a sketch. Label the physical dimensions.

V = volume

$$\frac{dr}{dt} = \text{change in radius} = 100 \text{ km/sec}$$

$$\frac{dV}{dt} = \text{change in volume} = ?$$

$d = 400{,}000 \text{ km} = 2r$

$r = 200{,}000 \text{ km} = \text{radius}$

(B) List all known variables and decide what the unknown is.

$$V = \frac{4}{3}\pi r^3$$

(C) The static equation is given that relates the volume and the radius.

$$\frac{dV}{dt} = \frac{4}{3}\pi 3r^2 \frac{dr}{dt}$$

$$= 4\pi r^2 \frac{dr}{dt}$$

(D) Find the "dynamic" equation by finding the implicit derivative with respect to time.

$$\frac{dV}{dt} = 4\pi(200{,}000 \text{ km})^2 \left(100 \frac{\text{km}}{\text{sec}}\right)$$

(E) Take a "snapshot" at the instant when the radius is 200,000 km.

$$\frac{dV}{dt} = \text{change in volume}$$

$$= 4(\pi)(200{,}000)^2(100)$$

$$= 16{,}000{,}000{,}000{,}000\,\pi \text{ cubic km/sec}$$

(F) We do not need to determine any more values. We can multiply the values to find the answer.

c)

(A) Draw a sketch and label. Vertical distance is y, horizontal distance is x, distance from rocket to observer is s.

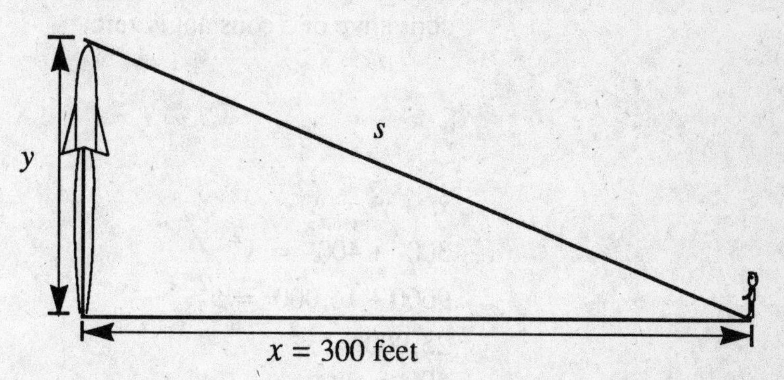

x = distance from launchpad to observer = 300 ft
y = height of rocket = 400 ft
s = line of sight distance from observer to rocket

(B) List all known values and decide what unknown is.

$$\frac{ds}{dt}$$

$$\frac{dy}{dt} = 20 \text{ ft/sec}$$

$$x^2 + y^2 = s^2$$

(C) Determine the static equation for this system. This equation follows from the Pythagorean formula for a right triangle.

$$2x\frac{dx}{dt} + 2y\frac{dy}{dt} = 2s\frac{ds}{dt}$$

$$x\frac{dx}{dt} + y\frac{dy}{dt} = s\frac{ds}{dt}$$

(divide by 2)

(D) Differentiate the static formula with respect to time. This is the dynamic equation.

$$(300)\left(\frac{dx}{dt}\right) + (400)\ (20) = s\left(\frac{ds}{dt}\right)$$

(E) Take a "snapshot" when the rocket is 400 feet high and is traveling at 20 feet/sec.

$$\frac{dx}{dt} = 0$$

because the distance from the launchpad and the observer does not change. It is constant and the derivative of a constant is zero.

From the substitutions in the dynamic equation, we can see that we are missing two values to solve for ds/dt. This often happens when we get to this step. However, $dx/dt = 0$ and the value of s can be found by returning to step (C).

$$x^2 + y^2 = s^2$$
$$300^2 + 400^2 = s^2$$
$$9000 + 16{,}000 = s^2$$
$$25{,}000 = s^2$$
$$500 = s$$

$$300\,(0) + 400\,(20) = 500\left(\frac{ds}{dt}\right)$$
$$0 + 8000 = 500\left(\frac{ds}{dt}\right)$$
$$\frac{8000}{500} = \frac{ds}{dt}$$
$$16 \text{ ft/sec} = \frac{ds}{dt}$$

(F) Now the value of $\frac{ds}{dt}$ can be found.

d)

(A) Draw a sketch and label.

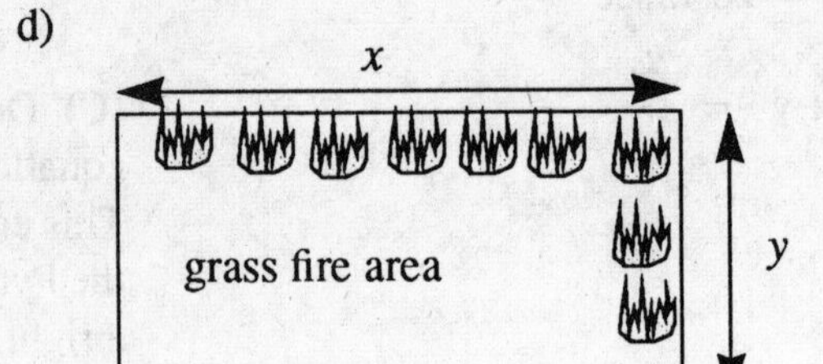

A = area of rectangle
x = N. boundary = 0.8 kilometers
y = E. boundary = 0.5 kilometers

$\frac{dy}{dt}$ = movement of *northern* boundary

= 0.3 km/hr

$\frac{dx}{dt}$ = movement of *eastern* boundary

= 0.2 km/hr

(B) All the known values are listed. The unknown is identified. Be careful that you note that movement *along* x is movement by the *eastern* boundary and movement *along* y is movement by the *northern* boundary.

$$\frac{dA}{dt} = \text{unknown}$$

$$A = x \cdot y$$

(C) Write the "static" equation. It is the formula for the area of the rectangle.

$$\frac{dA}{dt} = x\frac{dy}{dt} + y\frac{dx}{dt}$$

(use Product Rule)

(D) Differentiate the "static" equation with respect to time to find the "dynamic" equation.

$$\frac{dA}{dt} = x\frac{dy}{dt} + y\frac{dx}{dt}$$

$$= (0.8)(0.3) + (0.5)(0.2)$$

(E) Substitute the numbers into the "dynamic" equation.

$$\frac{dA}{dt} = (0.8)\,(0.3) + (0.5)\,(0.2)$$

$$= 0.24 + 0.10 = 0.34$$

(F) No other values are needed; therefore, the change in area can be found through multiplication and addition.

change in area = 0.34 sq.km/hr

5.5 ELASTICITY OF DEMAND

Normally, when an item increases in price, the demand for that item tends to fall. The increase in price may put this item beyond the ability for some to buy it. Conversely, more people can purchase an item if the price drops.

For a given item, the percent change in demand is reflected through the percent change in price. This variation changes from item to item. The equation for determining elasticity, E, for an item is based upon the ratio of percent change in demand versus percent change in price.

5.3 Elasticity of Demand

$$E = -\frac{p}{x} \cdot \frac{dx}{dp}$$

where E is the Elasticity of Demand, p is the price and x is the demand.

$\frac{dx}{dp}$ is the first derivative of the demand equation.

The value of E has three unique categories, $E < 1$, $E = 1$, $E > 1$. Table 5.2 outlines the meaning of each category.

E	Elasticity (Variability)	Meaning
< 1	inelastic demand	price changes do not affect demand much
$= 1$	unit elasticity	demand and price change by some amount
> 1	elastic demand	price changes greatly affect demand

Table 5.2 Value of E

Since price and demand are headed in opposite directions, $\frac{p}{x} \cdot \frac{dx}{dp}$ is always negative. Therefore, E is a positive quantity that measures the instantaneous responsiveness of demand to price.

EXAMPLE 5.9

a) Given the demand function $x = 750 - 25p$, find the elasticity of demand, E, when $p = 10$ and when $p = 20$.

b) The revenue function for an item is $R(x) = \frac{900}{x}$. Find the elasticity of demand, E, when $p = 16$.

SOLUTION 5.9

a) $x = 750 - 25p$ — Copy the demand function.

$$\frac{dx}{dp} = -25$$

Differentiate the demand function with respect to p.

$$E = -\frac{p}{x}\frac{dx}{dp}$$

The elasticity of demand formula is given.

$$E = \frac{-p}{750 - 25p} \cdot (-25) = \frac{-(-25)\,p}{750 - 25p}$$

Substitute for x and dx/dp into the elasticity of demand formula.

$$E = \frac{25p}{750 - 25p}$$

at p = 10:

The value of $p = 10$ is placed into the elasticity of demand formula with a result of $1/2$.

$$E = \frac{25\,(10)}{750 - 25\,(10)} = \frac{250}{750 - 250}$$

$$= \frac{250}{500} = \frac{1}{2}$$

$E < 1$; therefore, at $p = 10$, the system is relatively inelastic. A 20% *increase* in price will cause a 10% *decrease* in demand.

at p = 20:

The value of $p = 20$ is placed into the elasticity of demand formula. The result is 2.

$$E = \frac{25\,(20)}{750 - 25\,(20)} = \frac{500}{750 - 500}$$

$$= \frac{500}{250} = 2$$

$E > 1$; therefore, the system is relatively elastic at $p = 20$. A 10% *increase* in price will cause a 20% *decrease* in demand.

b) $R\,(x) = \dfrac{900}{x}$

The revenue function, $R\,(x)$, is given. The revenue function is given generally as $R\,(x) = xp$.

$$R\,(x) = \frac{900}{x} = xp$$

Find the demand function.

$$\frac{900}{x} = xp$$

therefore: $p = \dfrac{900}{x^2}$

$$p = \frac{900}{x^2}$$

Solve the demand function in terms of x.

$$x^2 p = 900$$

$$x^2 = \frac{900}{p}$$

$$x = \frac{30}{\sqrt{p}}$$

$$x = \frac{30}{\sqrt{p}} = 30p^{-1/2}$$

$$\frac{dx}{dp} = -\frac{1}{2}(30)p^{-3/2}$$

$$= -15p^{-3/2} = \frac{-15}{p^{3/2}}$$

$$= \frac{-15}{\sqrt{p^3}}$$

$$E = -\frac{p}{x}\frac{dx}{dp}$$

The elasticity of demand formula is given.

Since $x = \dfrac{30}{\sqrt{p}}, \dfrac{dx}{dp} = \dfrac{-15}{\sqrt{p^3}}$

$$E = \frac{-p}{\left(\dfrac{30}{\sqrt{p}}\right)} \frac{-15}{\sqrt{p^3}}$$

Substitute the known values into the elasticity of demand formula.

$$= \frac{15p}{\dfrac{30\sqrt{p^3}}{\sqrt{p}}} = \frac{15p}{30\sqrt{\dfrac{p^3}{p}}}$$

$$= \frac{15p}{30\sqrt{p^2}} = \frac{15p}{30p} = \frac{15}{30}$$

$$= \frac{1}{2}$$

The demand is fixed at $\frac{1}{2}$ for *all p*.

The demand is relatively inelastic at $1/2$. A 20% rise in price will cause a 10% decrease in demand.

5.6 ANNUITIES

The repayment of loans or the accumulation of savings can be done in payments called *annuities*. An *annuity* is a set of equal payments made at regular time intervals.

The periodic payments and the interest accumulated on these payments total to the *future value* or *amount* of the annuity. We can calculate the future value of an annuity by using the formula.

5.4 Annuity Formula

$$A = \frac{P\left[\left(1+\frac{r}{k}\right)^{kt} - 1\right]}{r/k}$$

A is the future value or amount.
P is the periodic payment.
r is the interest rate.
t is the time in years.
k is the number of payment periods per year.

When the annuity is compounded continuously, the formula is altered to one that contains the exponential, *e*.

5.5 Continuous Annuity Formula

$$A = \frac{P(e^{rt} - 1)}{r}$$

A = the amount at the end of a time period.
P = payment at the *beginning* of the period.
r = the interest rate for that period.
t = time period in years.
e = the natural numbers.

These types of formulas are covered in some calculus books.

EXAMPLE 5.10

a) What is the future value of an annuity with a periodic payment of $100, paid 4 times a year, at 8% interest for 5 years?

b) What is the rate of yearly increase at the second year for an annuity with 6% annual interest rate, compunded continuously and a payment of $1000?

SOLUTION 5.10

a) $P = \$100$ = periodic payment — List all of the known values.
$k = 4$ = number of annual payments
$r = 0.08$ = interest rate
t = time = 5 years

$$A = \frac{P\left[\left(1+\frac{r}{k}\right)^{kt} - 1\right]}{r/k}$$

The future value formula is used.

$$A = \frac{\$100\left[\left(1+\frac{0.08}{4}\right)^{4(5)} - 1\right]}{\frac{0.08}{4}}$$

Place the values into the formula.

$$A = \frac{\$100[(1+0.02)^{20} - 1]}{0.02}$$

Solve by combining and using order of operations.

$$A = \frac{\$100[(1.02)^{20} - 1]}{0.02}$$

$$A = \frac{\$100[1.48594 - 1]}{0.02}$$

Finding $(1.02)^{20}$ on a calculator requires the x^y key.

$$A = \frac{\$100[0.48594]}{0.02}$$

$$A = \$100(24.297)$$

$$A = \$2429.74$$

The future value is \$2429.74.

b) $$A = \frac{P(e^{rt} - 1)}{r}$$

The continuously compounded annuity formula is given.

$r = 0.06 =$ annual interest rate.
$p = \$1000 =$ periodic payment

$$\frac{dA}{dt} = ? \qquad t = 2 \text{ years}$$

$$A = \frac{1000(e^{0.06(t)} - 1)}{(0.06)}$$

Substitute the values into the given formula. t is not substituted for because we must first find the derivative with respect to 2.

$$A = 16{,}666.67\,(e^{0.6t} - 1)$$

$$A = 16{,}666.67\,e^{0.6t} - 16{,}666.67$$

$$\frac{dA}{dt} = 16{,}666.67\,e^{0.06t}(0.06) - 0$$

Find the first derivative.

$$\frac{dA}{dt} = 1000e^{0.06t}$$

$$\frac{dA}{dt} = 1000e^{0.06(2)}$$

Evaluate the derivative at $t = 2$.

$$\frac{dA}{dt} = 1000e^{0.12}$$

$$\frac{dA}{dt} = 1000(1.12749)$$

$$\frac{dA}{dt} = \$1127.49 \text{ (the rate the account is growing in the second year)}$$

5.8 EXPONENTIALS AND COMPOUND INTEREST

When the growth of an organism or a monetary account occurs *continuously,* the formula that often applies is one involving the base number, *e*.

5.6 Continuous Growth Formula

$$A = Pe^{rt}$$

A = amount at end of a time interval.
P = beginning amount.
r = growth (or interest) rate.
t = time interval.
e = natural number (approximately 2.71828).

With a slight modification, Formula 5.6 can be turned into a "decay" formula. A decay is essentially the opposite of growth. The formula for decay is:

5.7 Continuous Decay Formula

$$A = Pe^{-rt}$$

A = amount at end of a time interval.
P = beginning amount.
r = decay (or loss) rate.
t = time interval.
e = natural number (approximately 2.71828).

We can find the growth or decay for the *entire* time interval by using the basic Formulas 5.6 and 5.7. If we want to find the instantaneous rate of growth or decay at any *instant of time* within the interval, then we use the first derivative of Formula 5.6 or 5.7. The derivative is found *after* all known values have been substituted in the formulas.

EXAMPLE 5.11

a) What is the population of a city of 10,000 people after 10 years of growth? The growth rate is 5% a year.

b) How long does it take an account to double from $5000 to $10,000 if the interest rate of return is 8% a year?

c) A beaver grows according to rate, r, equal to 0.35 during its first 10 years of life. How fast is the beaver growing in its 6th year of life assuming a newborn beaver weighs 0.5 pound?

d) The present value of future money is the amount of money needed to be deposited now to have a certain amount available in the future. How much should be invested now at 6% compounded continuously so that in 10 years you have $5000 invested?

SOLUTION 5.11

a) $A = Pe^{rt}$ — The continuous growth formula is needed.

$P = 10{,}000$ people
$r = 5\% = 0.05$
$t = 10$ years

The values to be substituted into the formula are defined.

$A = 10{,}000\, e^{(0.05)\,(10)}$ — Substitute the values into the equation.

$A = 10{,}000\, e^{0.50}$

$A = 10{,}000\,(1.6487)$ — $e^{0.5} = 1.6487$

$A = 16{,}487$ people — The city has grown to 16,487 people in 10 years.

b) $A = Pe^{rt}$ — The amount "grows" in the account. The growth formula is needed.

$A = \$10{,}000$
$B = \$5000$
$r = 8\% = 0.08$

List the values given in the problem.

$\$10{,}000 = 5000e^{0.08t}$ — Substitute the given values into the formula.

$$\frac{10{,}000}{5000} = \frac{5000}{5000}e^{0.08t}$$

Divide both sides by 5000.

$2 = e^{0.08t}$

$\ln 2 = \ln e^{0.08t}$

Since the unknown value of t is in the exponent, the rules of logarithms are needed.

$\ln 2 = 0.08t \ln e$

The natural logarithm of both sides is used.

$\ln 2 = 0.08t(1)$

The natural logarithm of e is equal to 1 (see chapter 1).

$\ln 2 = 0.08t$

$\frac{\ln 2}{0.08} = t$

A logarithm table or a calculator tells us that $\ln 2 = 0.6931$.

$\frac{0.6931}{0.08} = t$

$8.66 = t$

doubling time = 8.66 years

The length of time needed to double the amount of money is about 8 2/3 years.

c) $A = Pe^{rt}$

The growth formula is needed to determine the growth rate of a beaver.

$\frac{dA}{dt} = \text{unknown rate}$

$P = 0.5$ pounds
$r = 0.35$
$t = 6$ years

List the values given in the problem.

$A = 0.5e^{(0.35)(t)}$

$A = 0.5e^{0.35t}$

Substitute the given value into the formula. Do not substitute for t until the formula is differentiated.

$\frac{dA}{dt} = 0.5e^{0.35t}(0.35)$

$\frac{dA}{dt} = 0.175e^{0.35t}$

Differentiate with respect to t.

$\frac{dA}{dt} = 0.175e^{0.35(6)}$

Substitute $t = 6$ into the derivative.

$\frac{dA}{dt} = 0.175e^{2.1}$

$\frac{dA}{dt} = 1.33 \text{ pounds/year}$

Solve to find the growth rate for the 6th year.

d) $A = Pe^{rt}$

The amount needed to invest is found through the decay formula because we allow the future amount to drop in value to the "present value."

$P = 5000$
$r = 6\% = 0.06$
$t = 10$ years

List the values from the problem.

$A = 5000e^{-0.06\,(10)}$

$A = 5000e^{-0.6}$

$A = 5000\,(0.5488)$

$A = \$2744.05$

Substitute the given values into the formula. *t* is also substituted since we have no derivative to find for this problem. We are not finding a *rate*. $e^{-0.6} = 0.5488$ (from a calculator or a table).

\$2744.05 must be invested to have \$5000 in the account in 10 years.

Practice Exercises

1. Find the marginal cost of producing the 5th unit, if the cost equation is $C(x) = 3x^2 + 14$.

2. Find the marginal profit of the 5th unit if $P(x) = 4\sqrt{x} + 2$.

3. The sales for a retail outlet is $p(x) = 2000 - x$, where x is the number of units sold and $p(x)$ is the price per unit. Find the marginal revenue at 1001 units sold.

4. The demand equation for small appliances is $p(x) = 28 - \frac{x}{80}$ and the cost of production is: $C(x) = 10{,}000 + 3x$. Find the marginal profit for the 31st unit.

5. A business needs to fence in a parking lot. If the business has 1600 feet of fencing, what is the maximum area it can enclose?

6. Solve.
   a) A spherical balloon is losing air at about 8 in^3 per minute. How fast is the diameter decreasing when the radius is 4 inches?
   b) A pebble tossed in a pond causes a wave to radiate from the point of impact at 3 feet/second. How fast is the area enclosed by the circular wave increasing when the radius is 10 feet?

7. Given the demand function $x = 400 - 40p$, find the elasticity of demand, E, when $p = 5$.

8. What is the future value of an annuity with a periodic payment $100, paid 12 times a year, at 6% interest compounded continuously for 5 years?

9. What is the amount in an account that has $1000 principal invested at 5% interest, compounded continuously for 6 years?

Answers

1. The marginal cost is approximately 90 dollars via first derivative, 27 dollars if we use the subtraction method.

2. $P'(4) = 1$ dollar

3. $R'(1000) = 0$

4. $P'(30) = 24.25$

5. 6400 sq. ft.

6. a) $\frac{1}{4\pi}$ inch/minute

   b) $60\pi \text{ft}^2/\text{sec}$

7. $E = 1$

8. $6977.00

9. $1349.86

6

Techniques of Integration

6.1 INTRODUCTION AND PURPOSE

In earlier chapters of this text we concentrated upon the process called "differentiation." The derivative helped us find values at a given instant in time or to get a point-by-point look at a function.

As with many other operations in mathematics, the derivative has a "reverse operation." This "reverse operation" is called *antidifferentiation* or *integration.* Instead of breaking things down into points or instants, the integration process sums or summarizes what a process does. In order to understand how the integral (or antiderivative) is formed, we will first cover sums in Sigma Notation and then cover antiderivative rules. More advanced techniques of integration follow.

The topics covered in this chapter are:

a) Summation notation
b) The indefinite integral
c) The definite integral
d) Integration by substitution
e) Integration by parts
f) Applications of integration

6.2 SUMMATION NOTATION

In calculus applications, the study of integration follows two seemingly unrelated paths. One path is antidifferentiation, the reversal of the rules for derivatives, and the other is summation, the process of combining several pieces together. In this section, we will briefly discuss the latter concept.

The Greek letter, Σ, is the summation symbol. The use of Σ means that several numbers (or terms) will be generated and then added.

For example:

$$\sum_{n=3}^{7} 4n$$

will generate a series of terms (numbers) to be added together. In our example, n is the summation index. The letters j, i, k, n, and m are often used to represent indices. The numbers 3 and 7 are the lower and upper limits of the index, respectively. The numbers substituted for the index in the expression following Σ are whole numbers from 3 to 7 inclusive. The expression, $4n$, changes with each new index number. This way a series of numbers is formed.

Our example of sigma notation,

$$\sum_{n=3}^{7} 4n$$

unfolds in a stepwise manner as follows:

$$\begin{aligned}
\text{Sum} &= \sum_{n=3}^{7} 4n \\
&= 4(3) + \sum_{n=4}^{7} 4n \\
&= 4(3) + 4(4) + \sum_{n=5}^{7} 4n \\
&= 4(3) + 4(4) + 4(5) + \sum_{n=6}^{7} 4n \\
&= 4(3) + 4(4) + 4(5) + 4(6) + \sum_{n=7}^{7} 4n \\
&= 4(3) + 4(4) + 4(5) + 4(6) + 4(7)
\end{aligned}$$

The "expansion process" of the symbolized sum is not usually done in a stepwise manner as shown above. The series is usually generated in one step. The method above was done for demonstration purposes.

The terms can now be multiplied and added to find the numerical sum.

$$4(3)+4(4)+4(5)+4(6)+4(7) = 12+16+20+24+28$$
$$= 100$$

Therefore, written in a shorthand manner:

$$\sum_{n=3}^{7} 4n = 100$$

The summation process will help us understand the symbol used in integration and it will prepare us for finding the area under a curve.

EXAMPLE 6.1

Find the following sums:

a) $\sum_{n=1}^{5} n^2$

b) $\sum_{k=2}^{5} 2^k$

c) $\sum_{k=1}^{4} k^2 - 2k$

d) $\sum_{i=1}^{7} 3$

e) $\sum_{j=5}^{9} 2x^j$

f) $\sum_{n=1}^{4} (-1)^n \left(\frac{n}{n+1}\right)$

SOLUTION 6.1

a) $\sum_{n=1}^{5} n^2$

Copy the given summation. This is the sum of n^2 as n goes from 1 to 5.

$$\sum_{n=1}^{5} n^2 = 1^2 + 2^2 + 3^2 + 4^2 + 5^2$$ Generate the series.

$$\sum_{n=1}^{5} n^2 = 1 + 4 + 9 + 16 + 25 = 55$$ The sum is 55.

b) $\sum_{k=2}^{5} 2^k$ Copy the given summation. This is the sum of 2^k as k goes from 2 to 5.

$$\sum_{k=2}^{5} 2^k = 2^2 + 2^3 + 2^4 + 2^5$$ Generate the series.

$$\sum_{k=2}^{5} 2^k = 4 + 8 + 16 + 32 = 60$$

c) $\sum_{k=1}^{4} k^2 - 2k$ Copy the given summation. This is the sum of $k^2 - 2k$ as k goes from 1 to 4.

$$\sum_{k=1}^{4} k^2 - 2k = (1^2 - 2(1)) + (2^2 - 2(2)) + (3^2 - 2(3)) + (4^2 - 2(4))$$ Generate the series.

$$\sum_{k=1}^{4} k^2 - 2k = (1-2) + (4-4) + (9-6) + (16-8)$$

$$= -1 + 0 + 3 + 8$$

$$= 10$$ The sum is 10.

d) $\sum_{i=1}^{7} 3$ Write the given summation. This is the sum of 3 as i goes from 1 to 7.

$$\sum_{i=1}^{7} 3 = 3 + 3 + 3 + 3 + 3 + 3 + 3$$ Note that the expression following the sigma does not contain an i, so the numbers 1 through 7 never actually enter into the sum. However 3 is added to itself seven times.

$$\sum_{i=1}^{7} 3 = 3+3+3+3+3+3+3 = 21$$

The result is the same as 7×3.

The previous example pointed to the dual nature of the index. The index (a) tells how many times the operation must be generated and (b) (normally) enters into the summation expression to generate a number or term.

e) $$\sum_{j=5}^{9} 2x^j$$

Copy the given summation. This is the sum of $2x^j$ as j goes from 5 to 9.

$$\sum_{j=5}^{9} 2x^j = 2x^5 + 2x^6 + 2x^7 + 2x^8 + 2x^9$$

Generate the series. Since we do not have numerical values for x, the summation is complete.

f) $$\sum_{n=1}^{4} (-1)^n \left(\frac{n}{n+1}\right)$$

Copy the given summation. This is the sum of $(-1)^n$ times $\left(\frac{n}{n+1}\right)$ as n goes from 1 to 4.

$$\sum_{n=1}^{4} (-1)^n \left(\frac{n}{n+1}\right) = \left[(-1)^1\left(\frac{1}{2}\right)\right] + \left[(-1)^2\left(\frac{2}{3}\right)\right] + \left[(-1)^3\left(\frac{3}{4}\right)\right] + \left[(-1)^4\left(\frac{4}{5}\right)\right]$$

Generate the series.

$$\sum_{n=1}^{4} (-1)^n \left(\frac{n}{n+1}\right) = (-1)\left(\frac{1}{2}\right) + (1)\left(\frac{2}{3}\right) + (-1)\left(\frac{3}{4}\right) + (1)\left(\frac{4}{5}\right)$$

$$= -\frac{1}{2} + \frac{2}{3} - \frac{3}{4} + \frac{4}{5}$$

Simplify the expression and combine fractions.

$$= -\frac{30}{60} + \frac{40}{60} - \frac{45}{60} + \frac{48}{60}$$

Determine that 60 is the least common denominator.

$$= \frac{-30 + 40 - 45 + 48}{60} = \frac{13}{60}$$ The numerators are combined.

$$\text{Therefore: } \sum_{n=1}^{4} (-1)^n \left(\frac{n}{n+1}\right) = \frac{13}{60}$$

Indefinite Integrals

We have discussed many rules for derivatives in earlier sections of this book. Three of these rules are shown in Table 6.1.

Original Function	Derivative
$f(x) = x^n$ where n is a *numerical* exponent.	$f'(x) = nx^{n-1}$
$f(x) = e^x$	$f'(x) = e^x$
$f(x) = \ln x$	$f'(x) = \frac{1}{x}$

Table 6.1 Rules for Derivatives

The three rules outlined in Table 6.1 are the basis for the discussion of *antiderivatives* in most Business Calculus texts. *All* of the rules for antiderivatives that follow will derive from these three rules for derivatives.

As you may already know, finding the antiderivative is the *reverse* of finding the derivative.

When we want to find the antiderivative (integral) of a function, we place that function *between* the symbols $\int$ and dx. The expression

$$\int f(x)\, dx$$

means "find the integral of $f(x)$." The origin of the use of these symbols will be discussed in the next section. Therefore, just as we use + for addition, we will use $\int dx$ for integration.

One other comment before we state the three basic rules for integration. You may have noticed that when two similar, yet different, functions like $f(x) = x^2 - 2$ and $f(x) = x^2 + 57$ are differentiated, the result for *both* is $f'(x) = 2x$.

When we desire to reverse the differentiation process, we will, unfortunately, have *no idea* what the constant was *before* differentiation. For this reason, $+C$ is added to *all indefinite integrals* because we do not *definitely* know what constant was present before differentiation.

The three primary rules for indefinite integrals are listed in Table 6.2.

Rule	Original Function	Integral	Result
I	$f(x) = x^n$, $n \neq -1$	$\int x^n\,dx$	$F(x) = \frac{x^{n+1}}{n+1} + C$
II	$f(x) = e^x$	$\int e^x\,dx$	$F(x) = e^x + C$
III	$f(x) = \frac{1}{x}$	$\int x^{-1}\,dx$	$F(x) = \ln\lvert x\rvert + C$

Table 6.2 Rules for Indefinite Integrals

Note: The expression $F(x)$ represents the integral of some function, $f(x)$.

Rule I from Table 6.2 is really the reverse of the Power Rule for derivatives. We just *add* one to the existing exponent and *divide* by the new sum. Rule II shows that the integral—like the derivative—of e^x is e^x. Rule III is for the special case when $n = -1$ for x^n. Since the derivative of $\ln x$ is $1/x$, the integral of $1/x$ is $\ln|x| + C$. The absolute value symbols are *absolutely necessary* because the logarithm of a negative quantity is undefined.

Before we begin an example of integrals, we note that:

$$\int k \cdot f(x)\,dx = k\int f(x)\,dx$$ Constant Multiple Rule

k any real number

$$\int (f(x) \pm g(x))\,dx = \int f(x)\,dx \pm \int g(x)\,dx$$ Sum & Difference Rule

EXAMPLE 6.2

Find the integral (antiderivative) of the following:

a) $f(x) = 2e^x$

b) $f(x) = 5x^2 + 2x$

c) $f(x) = \frac{5}{x}$

d) $4x^{3/2} - 2x^{1/2}$

e) $7e^x - \sqrt[3]{x}$

f) $f(x) = 6$

g) $f(w) = \dfrac{1}{w} - \dfrac{1}{\sqrt{w}}$

h) $f(t) = \dfrac{7-t^2}{t}$

i) $f(x) = \dfrac{x-1}{\sqrt{x}}$

SOLUTION 6.2

a) $f(x) = 2e^x$ — Copy the given function.

$\int 2e^x dx$ — Rewrite as an integral.

$2\int e^x dx$ — Move the constant in front of the integral symbol, $\int$.

$2e^x + C$ — Use Rule II to complete integration.

Note: Only numbers (or designated constants) can be moved in front of the $\int$ symbol.

Note: Many times students are concerned about what happens to the $\int$ and the *dx* when the integral is solved. These symbols indicate an operation to be performed and they "disappear" much like the plus sign in $2 + 2 = 4$.

b) $f(x) = 5x^2 + 2x$ — Copy the given function.

$\int (5x^2 + 2x)\, dx$ — Rewrite as an integral.

$\int 5x^2 dx + \int 2x dx$ — This integral can be separated at the plus sign.

$5\int x^2 dx + 2\int x^1 dx$ — Move the constants.

$\dfrac{5x^{2+1}}{2+1} + \dfrac{2x^{1+1}}{1+1} + C$ — Use Rule I to find the antiderivative (integral).

$\frac{5x^3}{3} + \frac{2x^2}{2} + C$	Combine numbers.
$\frac{5}{3}x^3 + x^2 + C$	Simplify the expression.
c) $f(x) = \frac{5}{x}$	Copy the given function.
$\int \frac{5}{x}\,dx$	Rewrite as an integral.
$5\int \frac{1}{x}\,dx$	Move the constant.
$5\ln\lvert x\rvert + C$	Use Rule III to solve.
d) $4x^{3/2} - 2x^{1/2}$	Copy the given function.
$\int (4x^{3/2} - 2x^{1/2})\,dx$	Rewrite as an integral.
$\int 4x^{3/2}dx - \int 2x^{1/2}dx$	Separate into two integrals.
$4\int x^{3/2}dx - 2\int x^{1/2}dx$	Move the constants.
$4\left(\frac{x^{3/2+1}}{\frac{3}{2}+1}\right) - 2\left(\frac{x^{1/2+1}}{\frac{1}{2}+1}\right) + C$	Solve using Rule I from Table 6.2.
$4\left(\frac{x^{3/2+2/2}}{\frac{3}{2}+\frac{2}{2}}\right) - 2\left(\frac{x^{1/2+2/2}}{\frac{1}{2}+\frac{2}{2}}\right) + C$	Simplify by combining constants.
$4\frac{x^{5/2}}{5/2} - 2\frac{x^{3/2}}{3/2} + C$	
$(4)\left(\frac{2}{5}\right)x^{5/2} - (2)\left(\frac{2}{3}\right)x^{3/2} + C$	Invert the fractions in the denominators.
$\frac{8}{5}x^{5/2} - \frac{4}{3}x^{3/2} + C$	Multiply to complete solution.
e) $7e^x - \sqrt[3]{x}$	Copy the given function.
$f(x) = 7e^x - x^{1/3}$	Convert the radical to a fractional exponent.

$\int (7e^x - x^{1/3})\,dx$	Rewrite as an integral.
$\int 7e^x dx - \int x^{1/3} dx$	Separate into two integrals.
$7\int e^x dx - \int x^{1/3} dx$	Move the constant.
$7e^x - \dfrac{x^{1/3+1}}{\frac{1}{3}+1} + C$	Evaluate using Rule II and Rule I from Table 6.2.
$7e^x - \dfrac{x^{1/3+3/3}}{\frac{1}{3}+\frac{3}{3}} + C$	
$7e^x - \dfrac{x^{4/3}}{4/3} + C$	Combine the constants in the exponent.
$7e^x - \frac{3}{4}x^{4/3} + C$	Invert the function in the denominator to obtain the solution.
f) $f(x) = 6$	Copy the given function.
$\int 6dx$	Rewrite as an integral.
$\int 6\,(1)\,dx$	Note that $6 = 6(1)$.
$\int 6x^0 dx$	$x^0 = 1$
$6\dfrac{x^{0+1}}{0+1} + C$	Use Rule I from Table 6.2 to evaluate the integral.
$6\dfrac{x^1}{1} + C$	Combine the constants.
$6x + C$	The solution has the net effect of placing a variable next to the constant.
g) $f(w) = \dfrac{1}{w} - \dfrac{1}{\sqrt{w}}$	Copy the given function.
$f(w) = \dfrac{1}{w} - \dfrac{1}{w^{1/2}}$	Convert the radical expression to a fractional exponent.

$f(w) = w^{-1} - w^{-1/2}$

$\int (w^{-1} - w^{-1/2})\, dw$ — Rewrite as an integral.

$\int w^{-1} dw - \int w^{-1/2} dw$ — Separate into two integrals.

$\ln|w| - \dfrac{w^{-1/2+1}}{-\frac{1}{2}+1} + C$ — Combine the constants.

$\ln|w| - \dfrac{w^{-1/2+2/2}}{-\frac{1}{2}+\frac{2}{2}} + C$

$\ln|w| - \dfrac{w^{1/2}}{\frac{1}{2}} + C$ — Invert the fraction in the denominator to obtain the final answer.

$\ln|w| - 2w^{1/2} + C$

h) $f(t) = \dfrac{7-t^2}{t}$ — Copy the given function.

$\int \dfrac{7-t^2}{t} dt$ — Rewrite as an integral.

$\int \dfrac{7}{t} - \dfrac{t^2}{t} dt$ — Since our rules do not cover rational expressions, we make use of division rules.

$\int (7t^{-1} - t)\, dt$

$\int 7t^{-1} dt - \int t dt$ — Separate the integral into two parts.

$7\int t^{-1} dt - \int t dt$ — Move constants.

$7\ln|t| - \dfrac{t^2}{2} + C$ — Use Rules III and I to find the antiderivative.

i) $f(x) = \dfrac{x-1}{\sqrt{x}}$ — Copy the function.

$f(x) = \dfrac{x-1}{x^{1/2}}$ — Convert the radical expression to a fractional exponent.

$\int \frac{x-1}{x^{1/2}} dx$	Rewrite as an integral.
$\int \left(\frac{x}{x^{1/2}} - \frac{1}{x^{1/2}} \right) dx$	Separate into two fractions.
$\int (x^{1-1/2} - x^{-1/2})\, dx$	Divide each fraction.
$\int x^{1/2} dx - \int x^{-1/2} dx$	Separate into two integrals.
$\frac{x^{3/2}}{\frac{3}{2}} - \frac{x^{1/2}}{\frac{1}{2}} + C$	Find the antiderivative of each.
$\frac{2}{3}x^{3/2} - 2x^{1/2} + C$	Invert each fraction to obtain the answer.

Evaluating Constants of Integration

Sometimes we are able to evaluate the constant of integration, C. This occurs when we know the value of the integral at a certain point.

EXAMPLE 6.3

Find the integral and the constant of integration for the following functions.

a) $f(x) = x+2, \quad F(0) = 7$

b) $f(x) = \sqrt{x}, \quad F(4) = 11$

c) $f(x) = \frac{4}{x}, \quad F(1) = 12$

d) $f(x) = e^{x} + 2, \quad F(0) = -7$

SOLUTION 6.3

a) $f(x) = x+2,\ F(0) = 7$	Copy the given function.
$\int (x+2)\, dx$	Rewrite as an integral.
$F(x) = \frac{x^2}{2} + 2x + C$	Evaluate the integral.

$F(0) = \frac{0^2}{2} + 2(0) + C = 7$	Equate the integrated function at 0 to the number 7.
$0 + C = 7$	
$C = 7$	The constant of integration is 7.
$F(x) = \frac{x^2}{2} + 2x + 7$	The complete antiderivative is shown.
b) $f(x) = \sqrt{x}, \quad F(4) = 11$	Copy the given function.
$f(x) = x^{1/2}$	Convert the radical to a fractional exponent.
$\int x^{1/2} dx$	Rewrite as an integral.
$F(x) = \frac{x^{3/2}}{\frac{3}{2}} + C = \frac{2}{3}x^{3/2} + C$	Evaluate the integral.
$F(4) = \frac{2}{3}(4)^{3/2} + C = 11$	$F(4) = 11$.
$\frac{2}{3}(\sqrt{4})^3 + C = 11$	
$\frac{2}{3}(2)^3 + C = 11$	Solve for C.
$\frac{2}{3}(8) + C = 11$	
$\frac{16}{3} + C = 11$	
$C = \frac{33}{3} - \frac{16}{3}$	Set $11 = \frac{33}{3}$.
$C = \frac{17}{3}$	The constant of integration is $\frac{17}{3}$.
$F(x) = \frac{2}{3}x^{3/2} + \frac{17}{3}$	Determine the complete integral.

c) $f(x) = \frac{4}{x}, \quad F(1) = 12$ — Copy the function.

$\int \frac{4}{x}\,dx$ — Rewrite as an integral.

$4\int \frac{1}{x}\,dx$ — Move the constant.

$F(x) = 4\ln|x| + C$ — Evaluate the indefinite integral.

$F(1) = 12 = 4\ln|1| + C$

$12 = 4(0) + C$ — $F(1) = 12$.

$12 = C$ — The constant of integration is 12.

$F(x) = 4\ln|x| + 12$ — Write the complete antiderivative.

d) $f(x) = e^x + 2, F(0) = -7$ — Copy the given function.

$\int (e^x + 2)\,dx$ — Rewrite as an integral.

$F(x) = e^x + 2x + C$ — Determine the indefinite integral.

$F(0) = e^0 + 2(0) + C = -7$

$1 + 0 + C = -7$

$C = -7 - 1$

$C = -8$ — The constant of integration is –8.

$F(x) = e^x + 2x - 8$ — Write the complete antiderivative.

6.3 THE DEFINITE INTEGRAL

We have talked about two of the uses of integration: (a) to reverse the differentiation process; and (b) to sum several elements. Perhaps, when you have been confronted with the task of finding the area of an irregularly shaped room or plot of land, you decided to break the area into little rectangles, triangles or circles. The total area could then be found by adding all the pieces together.

Consider the area bounded by the curve, $f(x)$, in Figure 6.1, vertical lines $x = a$, and $x = b$, and the x-axis.

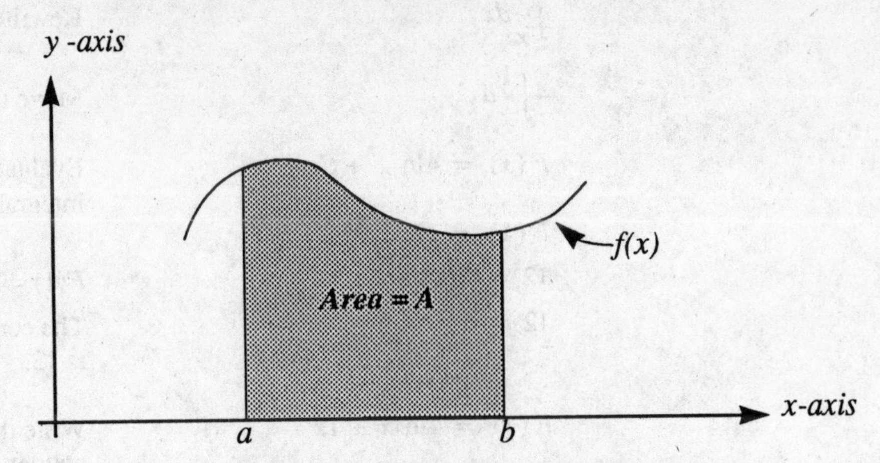

Figure 6.1 Area Under a Curve

The major problem in finding the area, A, is that the function, $f(x)$, forms a very irregular upper border. We can approximate the area by dividing it into several rectangles of *equal* width. We refer to the width as Δx (for change in x).

Figure 6.2 shows how the area might by divided.

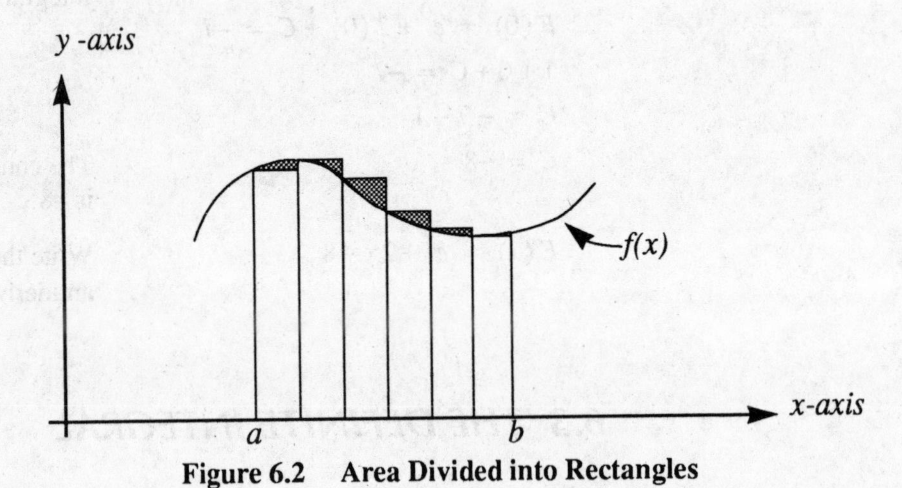

Figure 6.2 Area Divided into Rectangles

There are six rectangles that nearly sum to the area, A, except for the error indicated by the darkly shaded area in some rectangles. We have already stated that each rectangle is the *same width*. The height of each rectangle is determined by where we are on the curve, $f(x)$. Note that not all rectangles are the same height but have one top corner touching the curve. Figure 6.3 shows the close-up of one rectangle.

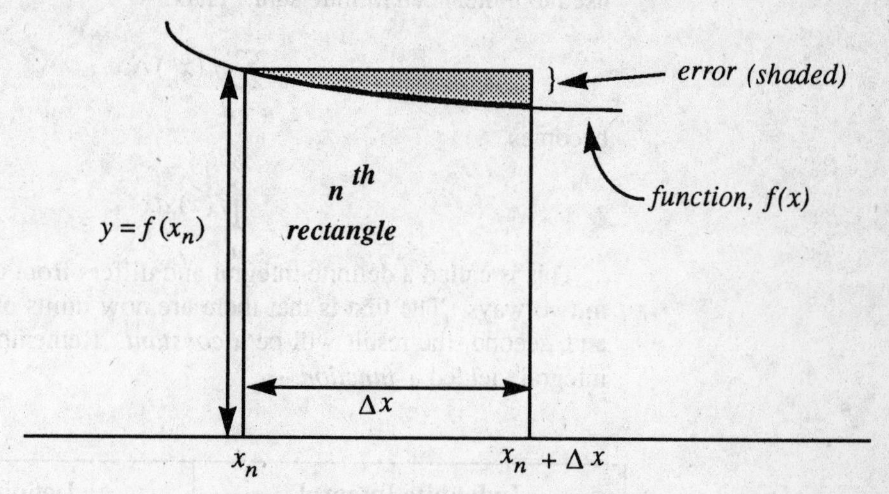

Figure 6.3 Close-Up of Rectangle

The area of a rectangle is known to be length times width, so for our rectangles of length, $f(x)$, and width, Δx, we have an area, $A = f(x) \cdot \Delta x$. If we use sigma notation to indicate a sum starting at a and ending at b, the total area is

$$\sum_{n=a}^{b} f(x_n)\,\Delta x.$$

$f(x_n)$ is a subscript notation, where n labels which height we are talking about.

Our error will diminish if we increase the number of rectangles, and make them very narrow. We will not have so much of the rectangle above or below the curve. See Figure 6.4.

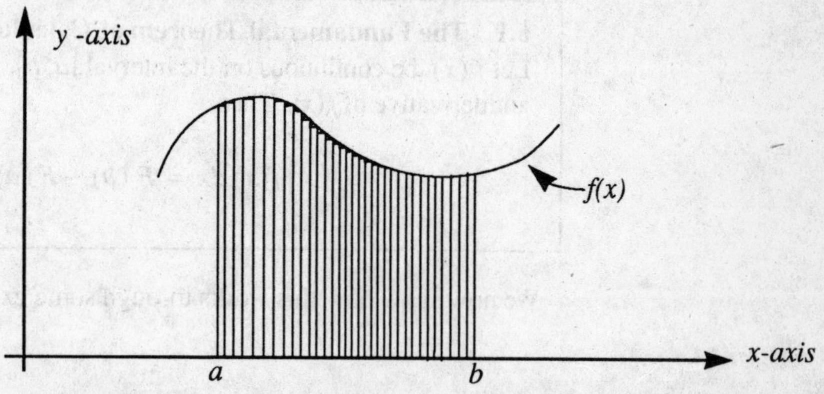

Figure 6.4 Approximating Area with Many Rectangles

Now if we allow our Δx to become infinitesimal (dx in Leibniz notation), we begin to see the reason for integral notation. The German, $\int$, is used to indicate an infinite sum. Thus,

$$\sum_{n=a}^{b} f(x_n)\,\Delta x$$

becomes

$$\int_{a}^{b} f(x)\,dx$$

This is called a definite integral and differs from the indefinite integral in two ways. The first is that there are now limits of a and b introduced, and, second, the result will be a *constant*. Remember that the indefinite integral yielded a *function*.

Indefinite Integral	Definite Integral
No limits of integration	Limits of integration
Result is a *function*	Result is a *constant*
$+C$ is needed	No $+C$ needed

Table 6.3 Differences Between an Indefinite Integral and a Definite Integral

The use of the definite integral gives rise to The Fundamental Theorem of Calculus, Formula 6.1.

6.1 The Fundamental Theorem of Calculus

Let $f(x)$ be continuous on the interval $[a, b]$, and let $F(x)$ be the antiderivative of $f(x)$. Then,

$$\int_{b}^{a} f(x)\,dx = F(b) - F(a)$$

We now show how this works through some examples.

EXAMPLE 6.4

Evaluate the following definite integrals.

a) $\int_{2}^{4} 2x dx$

b) $\int_{-1}^{3} (x^2 + 5x)\, dx$

c) $\int_{3}^{5} \frac{5}{x} dx$

d) $\int_{0}^{2} 2e^x dx$

SOLUTION 6.4

a) $\int_{2}^{4} 2x dx$ — Copy the integral.

$2\int_{2}^{4} x dx$ — Move the constant.

$\left.\frac{2x^2}{2}\right|_2^4$ — Find the antiderivative and retain the limits using a vertical bar. The $\int$ and dx are no longer needed.

$\left.\frac{2x^2}{2}\right|_2^4$ — Cancel the 2's.

$\left.x^2\right|_2^4 = (4)^2 - (2)^2$ — $F(4) - F(2)$.

$= 16 - 4 = 12$ — The value of the definite integral is 12.

Note: The antiderivative is evaluated by computing the *top number* (4, here) *minus the bottom number* (2, here).

Note: Many students ask, "What happened to the + C?" If we look at the previous example, we note the $F(x) = x^2 + C$ (if we had used the indefinite integral). Now,

$$F(4) = 4^2 + C \qquad \text{and} \qquad F(2) = 2^2 + C$$
$$= 16 + C \qquad\qquad\qquad = 4 + C$$

$$F(4) - F(2) = (16 + C) - (4 + C) = 16 + C - 4 - C$$
$$= 16 - 4 + 0$$ The C's subtract out!
$$= 12$$

The C's *always subtract out*, so we don't bother with them.

b) $\int_{-1}^{3} (x^2 + 5x)\,dx$ — Copy the integral.

$\left.\frac{x^3}{3} + \frac{5x^2}{2}\right|_{-1}^{3}$ — Find the antiderivative and retain limits.

$F(3) \quad - \quad F(-1)$

$\left[\left(\frac{3^3}{3} + \frac{5(3)^2}{2}\right) - \left(\frac{(-1)^3}{3} + \frac{5(-1)^2}{2}\right)\right]$ — Evaluate function at both limits.

$\left(\frac{27}{3} + \frac{5\cdot 9}{2}\right) - \left(\frac{-1}{3} + \frac{5\cdot 1}{2}\right)$ — Use order of operations to simplify.

$\left(9 + \frac{45}{2}\right) - \left(-\frac{1}{3} + \frac{5}{2}\right)$

$\left(\frac{18}{2} + \frac{45}{2}\right) - \left(-\frac{2}{6} + \frac{15}{6}\right)$ — Find least common denominators.

$\frac{63}{2} - \frac{13}{6}$

$\frac{189}{6} - \frac{13}{6} = \frac{176}{6}$ — The value of the definite integral is $176/6$.

c) $\int_{3}^{5} \frac{5}{x}\,dx$	Copy the integral.
$5\int_{3}^{5} \frac{1}{x}\,dx$	Move the constant.
$5\ln\lvert x\rvert \Big|_{3}^{5}$	The antiderivative is the natural logarithm.
$F(5) \quad - \quad F(3)$ $5\ln\lvert 5\rvert - 5\ln\lvert 3\rvert$	Evaluate at the two limits.
$5\,(\ln 5 - \ln 3)$	The 5 is factored out.
$5\left(\ln\frac{5}{3}\right)$	The rules of logarithms make the fraction $\frac{5}{3}$.
$5\left(\ln\frac{5}{3}\right)$ $5\,(0.511) \cong 2.555$	We can leave this as the final answer, or we can approximate using decimals.
d) $\int_{0}^{2} 2e^{x}\,dx$	Copy the integral.
$2\int_{0}^{2} e^{x}\,dx$	Move the constant.
$2e^{x}\Big|_{0}^{2}$	Find the antiderivative and retain the limits.
$F(2) \quad - \quad F(0)$ $2e^{2} - 2e^{0}$ $2e^{2} - 2\,(1)$ $2e^{2} - 2$	Evaluate the integral at the limits and subtract.
$\cong 2\,(7.389) - 2$ $\cong 15.778 - 2 \cong 13.778$	The answer can be left in this form or approximated using decimals.

6.4 INTEGRATION THROUGH SUBSTITUTION (u-SUBSTITUTION)

The Chain Rule for derivatives was discussed in Chapter 3. The following three examples show some results of that rule.

Original Function	Derivative
$f(x) = (x^2 - 5x)^3$	$f'(x) = 3(x^2 - 5x)^2 (2x - 5)$
$f(x) = e^{x^2 - 1}$	$f'(x) = e^{x^2 - 1}(2x)$
$f(x) = \ln(x^2 - 2)$	$f'(x) = \frac{1}{x^2 - 2}(2x) = \frac{2x}{x^2 - 2}$

Table 6.4 Chain Rule Applications

Whereas the original function can be viewed as a single function, the derivative can be viewed as a product or a quotient of more than one function. Knowledge of the Chain Rule will help with the integration of more complex functions.

We can, sometimes, deduce the proper format necessary to allow a product or quotient of functions to be integrated. This can be done by letting some *part* of the integrand (function to be integrated) equal a single variable. By convention, the variable, u, is used for the substitution variable.

Table 6.2 is adapted in Table 6.5 to show how the basic three rules are modified for use with the substitution variable.

Rule	Original Function	Integral	Result
I	$f(u) = u^n$, $n \neq 1$	$\int u^n du$	$\frac{u^{n+1}}{n+1} + C$
II	$f(u) = e^u$	$\int e^u du$	$e^u + C$
III	$f(u) = \frac{1}{u}$	$\int u^{-1} du$	$\ln\lvert u \rvert + C$

Table 6.5 Integration Rules Adapted for u-Substitution

The difference between Table 6.5 and Table 6.2 is that u stands for a polynomial, exponential, or logarithmic function. For example, u can equal $3x^3 - 2x - 1$ or e^{2x} or $\ln(x^2 - 1)$ or any other similar function. Many students feel that the selection of the proper u-substitution can be difficult. This selection process is not exact, and *must be done through trial and error.*

After a demonstration of u-substitution, a list of *suggestions* will be presented to help facilitate the substitution process.

The integral of $f(x) = 2x(x^2+6)^5$ can be found by using Rule I in Table 6.5. We assume that this function could be the result of a Chain Rule operation for derivatives. The task at hand, then, is to find

$$\int 2x(x^2+6)^5\,dx$$

by converting to an integral in terms of u. In order to establish a pattern we let $u = x^2 + 6$ (*not* $(x^2+6)^5$).

If

$$u = x^2 + 6$$

then

$$\frac{du}{dx} = 2x$$

In Leibniz notation, $\frac{du}{dx}$ can be treated as a fraction, thus:

$$du = 2xdx$$

If the original integral is rewritten as

$$\int \underbrace{(x^2+6)}_{u}{}^5\, \underbrace{2xdx}_{du}$$

the substituion of u and du becomes apparent, and the integral now is

$$\int (u)^5 du$$

which can be easily integrated using Rule I from Table 6.5:

$$\frac{u^6}{6} + C$$

Since u was originally equated to $x^2 + 6$, the answer is

$$\int 2x(x^2+6)^5\,dx = \frac{(x^2+6)^6}{6} + C$$

It is confusing to find a derivative within the integration process. Remember that the differentiation is used *only to find a pattern* and should be considered as "scratchwork" done in conjunction with the integration.

The example in this text follows only the three rules presented in Table 6.5. Some suggestions for u-substitutions follow.

SUGGESTIONS FOR u-SUBSTITUTION

a) If the integrand (the function to be integrated) contains a polynomial within parentheses carried to a power, let u equal the contents of the parentheses.

$$\text{for } 3x^2(x^3-8)^5 \qquad \text{let } u = x^3 - 8$$

b) If the integrand is a radical expression, let u equal the contents of the radical.

$$\text{for } 5\sqrt{5x-2} \qquad \text{let } u = 5x-2$$

c) If the integrand contains e with an exponent other than x, let u equal the exponent.

$$\text{for } e^{x^2-2} \qquad \text{let } u = x^2 - 2$$

d) If the integrand is a fraction, let u equal the denominator of the fraction, excluding any numerical powers of the denominator.

$$\text{for } \frac{12x-5}{6x^2-5x} \qquad \text{let } u = 6x^2 - 5x$$

$$\text{for } \frac{4}{(2x+1)^3} \qquad \text{let } u = 2x+1$$

Finally, it is again important to remember that only constants can be "moved across" the $\int$ symbol. A constant may be needed to complete some substitution problems. If a constant is appended to the integrand, its *reciprocal must be placed* in front of the $\int$ symbol. The net effect is as if the integrand were multiplied by 1.

EXAMPLE 6.5

Find the following integrals through substitution.

a) $\int 3x^2(x^3+7)^6\,dx$

b) $\int x(3x^2+1)^5\,dx$

c) $\int x\sqrt{x^2+4}\,dx$

d) $\int 8xe^{5x^2-1}\,dx$

e) $\int \frac{2x-5}{x^2-5x+1}\,dx$

f) $\int \frac{x}{\sqrt{x^2+3}} dx$

g) $\int \frac{\ln x}{x} dx$

h) $\int \frac{4}{x+5} dx$

SOLUTION 6.5

a) $\int 3x^2 (x^3+7)^6 dx$	Copy the integral.
$\int 3x^2 \cdot (x^3+7)^6 dx$	This is the product of two functions, $3x^2$ and $(x^3+7)^6$.
Scratchwork:	
$u = x^3+7$	Let $u = x^3+7$.
$\frac{du}{dx} = 3x^2$	
$du = 3x^2 dx$	We need $3x^2 dx$ to obtain du.
$\int \underbrace{(x^3+7)}_{u}{}^6 \underbrace{3x^2 dx}_{du}$	Rewrite the integral to facilitate substitution.
$\int u^6 du$	The format follows Rule I of Table 6.5.
$\frac{u^7}{7} + C$	Find the indefinite integral.
$\frac{(x^3+7)^7}{7} + C$	Replace x^3+7 for u to get final result.

The result can be checked through differentiation.

b) $\int x(3x^2+1)^5 dx$	Copy the integral. This integral is the product of x and $(3x^2+1)^5$.
Scratchwork:	
$u = 3x^2+1$	Let $u = 3x^2+1$.
$\frac{du}{dx} = 6x$	

$du = 6xdx$	We need $6xdx$ to obtain du.
$\int (3x^2+1)^5 xdx$	Rewrite the integral.
$\frac{1}{6}\int \underbrace{(3x^2+1)}_{u}^5 \underbrace{6xdx}_{du}$	The given integral does not have $6xdx$, but we can add the constant by "counter-balancing" by the reciprocal of 6, $1/6$.
$\frac{1}{6}\int u^5 du$	The integral now fits the format of Rule I, Table 6.5.
$\frac{1}{6}\left(\frac{u^6}{6}\right)+C$	Find the indefinite integral.
$\frac{u^6}{36}+C$	Multiply fractions.
$\frac{(3x^2+1)^6}{36}+C$	Substitute for u to obtain final answer.

Check through differentiation.

c) $\int x\sqrt{x^2+4}dx$	Copy the given integral.
$\int x(x^2+4)^{1/2} dx$	Rewrite the radical as a fractional exponent.
Scratchwork:	The integrand can be viewed as the product of two functions.
$u = x^2+4$	Let $u = x^2+4$.
$\frac{du}{dx} = 2x$	We need $2xdx$ to obtain a du.
$du = 2xdx$	
$\int (x^2+4)^{1/2} xdx$	Rewrite the integral.
$\frac{1}{2}\int \underbrace{(x^2+4)}_{u}^{1/2} \underbrace{2xdx}_{du}$	A 2 can be included in the integrand if a $1/2$ is placed before the $\int$ symbol.
$\frac{1}{2}\int u^{1/2} du$	The format follows Rule I of Table 6.5.

$\frac{1}{2}\left(\frac{u^{3/2}}{3/2}\right)+C$	Find the indefinite integral.
$\frac{1}{2}\cdot\frac{2}{3}\cdot(u^{3/2})+C$	Simplify the form.
$\frac{1}{3}u^{3/2}+C$	
$\frac{1}{3}(x^2+4)^{3/2}+C$	Replace u with x^2+4.

Check the result through differentiation.

d) $\int 8xe^{5x^2-1}dx$	Copy the given integral.
$\int 8xe^{5x^2-1}dx$	The integrand is the product of $8x$ and e^{5x^2-1}.

Scratchwork:

$u = 5x^2-1$ $\frac{du}{dx} = 10x$ $du = 10xdx$	Substitution of u for the exponent of e is a suggested start.
$\int e^{5x^2-1}8xdx$	Rewrite the order of the integrand.
$8\int e^{5x^2-1}xdx$	Our scratch work calls for $10x$ not $8x$. Place the 8 before the $\int$ symbol.
$8\cdot\frac{1}{10}\int e^{\overbrace{5x^2-1}^{u}}\underbrace{10xdx}_{du}$	The constant, 10, is placed before the xdx and $1/10$ is multiplied by 8.
$\frac{8}{10}\int e^u du$	The format follows Rule II of Table 6.5.
$\frac{4}{5}e^u+C$	The antiderivative is determined. ($8/10$ was reduced to $4/5$).
$\frac{4}{5}e^{5x^2-1}+C$	Let $u = 5x^2-1$ to obtain the final result.

The result can be checked through differentiation.

e) $\int \frac{2x-5}{x^2-5x+1}dx$	Copy the given integral.
Scratchwork:	
$u = x^2-5x+1$ $\frac{du}{dx} = 2x-5$	The suggested technique is to let u equal the denominator, x^2-5x+1.
$du = (2x-5)\,dx$	We need $(2x-5)dx$ to obtain du.
$\int \underbrace{\frac{1}{x^2-5x+1}}_{u} \underbrace{(2x-5)\,dx}_{du}$	Rewrite the integral.
$\int \frac{1}{u}du$	The format follows Rule III of Table 6.5.
$\ln\lvert u\rvert + C$	The antiderivative is determined.
$\ln\lvert x^2-5x+1\rvert + C$	Let $u = x^2-5x+1$ to obtain the final result.

Check the result through differentiation.

f) $\int \frac{x}{\sqrt{x^2+3}}dx$	Copy the integral.
Scratchwork:	
$u = x^2+3$ $\frac{du}{dx} = 2x$	The suggested technique is to let the u equal the contents of the radical in the denominator.
$du = 2xdx$	We need $2xdx$ to obtain du.
$\int \frac{x}{(x^2+3)^{1/2}}dx$	Replace the radical sign with a fractional exponent.
$\int (x^2+3)^{-1/2}\,xdx$	Convert the exponent in the denominator to a negative exponent.
$\frac{1}{2}\int (x^2+3)^{-1/2}\,2xdx$	We need $2xdx$, so place a 2 in the integrand and place 1/2 before $\int$.

$\frac{1}{2}\int \underbrace{(x^2+3)^{-1/2}}_{u} \underbrace{2x\,dx}_{du}$ — Rewrite the integral.

$\frac{1}{2}\int u^{-1/2}\,du$ — The form is that found in Rule I of Table 6.5.

$\frac{1}{2}\left(\frac{u^{1/2}}{1/2}\right)+C$ — Find the integral.

$\frac{1}{2}\cdot\frac{2}{1}(x^2+3)^{1/2}+C$ — Replace u with x^2+3 to obtain the final result.

$(x^2+3)^{1/2}+C$

g) $\int \frac{\ln x}{x}\,dx$ — Copy the integral.

Scratchwork: (Trial 1)

$u = x$ — Normally, we might let u equal the denominator, x.

$\frac{du}{dx} = 1$

$du = dx$ — Note that the value of du differs from $\ln x\,dx$ by *more than just a constant.* Therefore, this choice of u-substitution is wrong.

Scratchwork: (Trial 2)

$u = \ln x$

$\frac{du}{dx} = \frac{1}{x}$

$du = \frac{1}{x}dx$ — Our second trial has $du = \frac{1}{x}dx$.

$\int \underbrace{(\ln x)}_{u} \underbrace{\frac{1}{x}dx}_{du}$ — Rewrite the integral.

$\int u\,du$ — The format is that of Rule I, Table 6.5.

$\frac{u^2}{2}+C$ — Find the antiderivative.

$\frac{(\ln x)^2}{2} + C$ — $u = \ln x$

Check the final result through differentiation. — Our example of a wrong u-substitution is typical of this trial-and-error process.

h) $\int \frac{4}{x+5} dx$ — Copy the integral.

$4\int \frac{1}{x+5} dx$ — Move the constant.

Scratchwork:

$u = x+5$ — Let $u = x + 5$.

$\frac{du}{dx} = 1$

$du = dx$ — Our du value must equal dx.

$4\int \underbrace{\frac{1}{x+5}}_{u} \underbrace{dx}_{du}$ — Rewrite the integral.

$4\int \frac{1}{u} du$ — The integrand follows the format of Rule III in Table 6.5.

$4\ln|u| + C$

$4\ln|x+5| + C$ — Our substitution was $u = x + 5$.

Check the result through differentiation.

Some Integrals Solved with Algebra

The process of u-substitution is very useful, but sometimes a simple algebraic operation will work instead. If several attempts at substitution solutions fail, try looking for an algebraic solution. The next example shows how some of these solutions work.

EXAMPLE 6.6

Find the following integrals:

a) $\int (x+2)\,(x+3)\,dx$

b) $\int \left(\frac{x^2 - 3x - 10}{x-5}\right) dx$

c) $\int \left(\frac{x}{x+7} + \frac{7}{x+7}\right) dx$

d) $\int \frac{e^{2x}}{e^x}\,dx$

SOLUTION 6.6

a) $\int (x+2)\,(x+3)\,dx$

Copy the integral.

$u = x+2 \quad or \quad u = x+3$

$\frac{du}{dx} = 1 \qquad \frac{du}{dx} = 1$

$du = dx \qquad du = dx$

u-substitution for either of the binomials gives only a $du = dx$.

$\int \underbrace{(\ x+2\ )}_{u}\ \underbrace{(\ x+3\ )}_{?}\ \underbrace{dx}_{du}$

or

$\int \underbrace{(\ x+2\ )}_{?}\ \underbrace{(\ x+3\ )}_{u}\ \underbrace{dx}_{du}$

We must have more than $dx = du$. Only *one* of the binomials can be accounted for by a *u*-substitution. du can differ from dx by only a constant.

$\int (x^2 + 5x + 6)\,dx$

We can, however, multiply the two binomials.

$\frac{x^3}{3} + \frac{5x^2}{2} + 6x + C$

The integral can be found through successive applications of the Power Rule. (Rule I, Table 6.2.)

b) $\int \left(\frac{x^2\ -\ 3x\ -\ 10}{x-5}\right) dx$

Copy the given integral.

$u\ =\ x^2 - 3x - 10 \quad or \quad u\ =\ x - 5$

$\frac{du}{dx}\ =\ 2x - 3 \qquad \frac{du}{dx}\ =\ 1$

$du\ =\ (2x-3)\,dx \qquad du\ =\ dx$

Letting $u\ =\ x^2 - 3x - 10$ or $u = 5 - x$ does not give the other polynomial as part of a *du*-value.

$\int \frac{\overbrace{x^2 - 3x - 10}^{u}}{\underbrace{x-5}_{?}}\ \underbrace{dx}_{?}$

or

There is no $2x - 3$ present for one case and more than dx present in the second case.

$$\int \overbrace{\underbrace{\frac{x^2 - 3x - 10}{x - 5}}_{du}}^{?} \underbrace{dx}_{du}$$

$$\int \frac{(x-5)(x+2)}{(x-5)} dx$$

Our last solution is to factor $x^2 - 3x - 10$ and cancel.

$$\int (x+2)\, dx$$

This produces an integrand that is easily solved using Rule I, Table 6.2.

$$\frac{x^2}{2} + 2x + C$$

c) $\displaystyle\int \left(\frac{x}{x+7} + \frac{7}{x+7} \right) dx$

Copy the given integral.

$u = x + 7$

$\dfrac{du}{dx} = 1$

$du = dx$

Letting $u = x + 7$ is suitable for one fraction but not the other.

$$\int \left(\frac{\overbrace{x}^{?}}{\underbrace{x+7}_{u}} + \frac{7}{\underbrace{x+7}_{u}} \right) \underbrace{dx}_{du}$$

$$\int \left(\frac{x}{x+7} + \frac{7}{x+7} \right) dx$$

$$\int \frac{x+7}{x+7} dx$$

$$\int 1 dx$$

We note that both fractions have a common denominator. We can combine them and simplify.

$$x + C$$

This simplification yields a very easy integral, solved by using Rule I, Table 6.2.

d) $\int \frac{e^{2x}}{e^x} dx$ Copy the given integral.

$\int \frac{e^x \cdot e^x}{e^x} dx$ One aspect of the integrand is a bit subtle. That is that $e^{2x} = e^x \cdot e^x$.

$\int \frac{e^x \cdot e^x}{e^x} dx$ Once we know the replacement for e^{2x}, we can cancel.

$\int e^x$ The result is Rule II, Table 6.2.

$e^x + C$ The use of exponents gives us a simple result.

6.5 INTEGRATION BY PARTS

The evaluation of certain integrals cannot be done using the methods discussed so far in this chapter. One additional technique that may work is Integration by Parts. This process often converts more complicated integrals to less complex forms that can be integrated more easily.

The Product Rule of differentiation lies behind the Integration by Parts method. Formula 6.2 shows the Product Rule of Derivatives in Leibniz form.

6.2 Product Rule of Derivatives

$$\frac{d(u \cdot v)}{dx} = u\frac{dv}{dx} + v\frac{du}{dx}$$

where u and v are differentiable functions.

As with Leibniz notation in many other areas, the dx can be treated as a fraction denominator. We can multiply each term by dx to clear functions.

$$d(u \cdot v) = udv + vdu$$

If we integrate both sides of the equation:

$$\int d(u \cdot v) = \int udv + \int vdu$$

$$u \cdot v = \int udv + \int vdu$$

If $\int v\,du$ is subtracted from both sides:

$$u \cdot v - \int v\,du = \int u\,dv$$

This can be switched around the equal sign to obtain Formula 6.3.

> **6.3 Integration by Parts**
>
> $$\int u\,dv = u \cdot v - \int v\,du$$
>
> if u and v are differentiable functions.

When Integration by Parts is used, the integrand must be separated into two functions u and v. This separation should be done such that a simpler integral is produced.

Consider

$$\int xe^{3x}\,dx$$

The substitution process does *not* work since letting $u = 3x$ will become $du = 3dx$. The other x in the integrand is not accounted for.

We must predict the best match of $\int u\,dv$ for $\int xe^{3x}\,dx$ (which is not always easy). We want our selection to be such that $\int v\,du$ is the *simplest* possible.

It is best to substitute $dv = e^{3x}dx$ and $u = x$ thus, because if $u = x$, then $du = dx$. The value of v is found through

$$\int dv = \int e^{3x}dx = \frac{1}{3}e^{3x}$$

(C is ignored until the end of the problem). In summary, we have

$$v = \frac{1}{3}e^{3x} \qquad\qquad u = x$$

$$dv = e^{3x}dx \qquad\qquad du = dx$$

Formula 6.3 can be rewritten from

$$\int u\,dv = uv - \int v\,du$$

to:

$$\int \underbrace{x}_{u}\ \underbrace{e^{3x}dx}_{dv} = \underbrace{x}_{u}\underbrace{\left(\frac{1}{3}e^{3x}\right)}_{v} - \int \underbrace{\frac{1}{3}e^{3x}}_{v}\ \underbrace{dx}_{du}$$

The *easier* integral of $\int \frac{1}{3} e^{3x}\,dx$ is now evaluated:

$$\int \frac{1}{3} e^{3x}\,dx = \frac{1}{9} e^{3x} + C$$

The overall solution, then, is

$$\int x e^{3x}\,dx = \frac{1}{3} x e^{3x} - \frac{1}{9} e^{3x} + C$$

EXAMPLE 6.7

Use Integration by Parts to solve the following integrals.

a) $\int \ln x\,dx$

b) $\int (x+1)\,e^{x}\,dx$

SOLUTION 6.7

a) $\int \ln x\,dx$ — Copy the given integral.

$\int u\,dv = \int \ln x\,dx$ — We want to obtain the proper selection for $\int u\,dv$.

$u = \ln x, \qquad v = x$

$\frac{du}{dx} = \frac{1}{x} \qquad dv = dx$

$du = \frac{1}{x}$

Since we do not know the integral of $\ln x$ (we know that $\int \frac{1}{x} dx = \ln|x|$ but not $\int \ln x dx$), we choose $u = \ln x$ and $dv = dx$. Make a table for *du, u, dv,* and *v.*

$$\int u\,dv = uv - \int v\,du$$

$$\int \ln x\,dx = \underbrace{\ln x}_{u} \cdot \underbrace{x}_{v} - \int \underbrace{x}_{v} \cdot \underbrace{\frac{1}{x}dx}_{du}$$

Match the values for the Formula 6.3.

$$= x\ln x - \int \frac{x}{x} dx$$

$$= x\ln x - \int 1\,dx$$

The right-hand side of the equal sign is simplified.

$$= x\ln x - x + C$$

The easier integral, $\int 1\,dx$, is solved to complete the problem.

$$\int \ln x dx = x\ln x - x + C$$

Check through differentiation.

b) $\int (x+1)\, e^x dx$ — Copy the given integral.

$\int u dv$ — We want to obtain the proper selection for $\int u dv$.

$u = x+1 \qquad v = e^x$ — The simplest substitution gives a table for u, du, v and dv.

$\frac{du}{dx} = 1 \qquad dv = e^x dx$

$du = dx$

$\int u dv = uv - \int v du$ — Substitute the table values into Formula 6.3.

$$\int \underbrace{(\ x+1\ )}_{u}\ \underbrace{e^x dx}_{dv} = \underbrace{(\ x+1\ )}_{u}\underbrace{e^x}_{v} - \int \underbrace{e^x}_{v}\underbrace{dx}_{dv}$$

$\int e^x dx = e^x + C$ — The simpler integral $\int e^x dx$ is found.

$$\int (x+1)\, e^x dx = (x+1)\, e^x - e^x + C$$ — This can be simplified.

$$= xe^x + e^x - e^x + C$$

$$= xe^x + C$$

Check the result through differentiation.

6.6 APPLICATIONS OF INTEGRATION

The applications discussed in this section fall into two areas (of many possible):

a) Area under the curve or total consumption
b) Consumer surplus

Area and Total Consumption

The evaluation of the definite integral is the procedure needed for solving this type of problem. The evaluation of the area or of the total consumption is based on the region bounded by the limits of integration, the x-axis, and the function under consideration.

EXAMPLE 6.8

a) Find the area bounded above by the function $f(x) = 2x$, below by the x-axis, and lying between $x = 0$ and $x = 4$.

b) Find the total area between $f(x) = x^2 - 4$ and the x-axis from $x = 0$ to $x = 3$.

c) The rate of consumption of oil by a small country is $C'(t) = 1.5e^{0.10t}$. The consumption is in millions of barrels and t is in years. How much oil does the country consume in 10 years?

SOLUTION 6.8

a)

A drawing of the region in question is very useful.

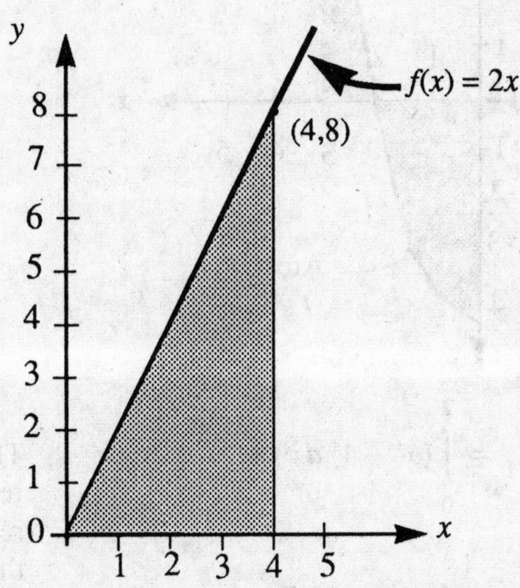

$$A = \int_0^4 2x\,dx$$

We are summing an infinite number of rectangular areas from $x = 0$ to $x = 4$. Thus, an integral is needed. (A = area)

$$A = x^2\Big|_0^4$$

$$A = 4^2 - 0^2 = 16 - 0$$

$$A = 16$$

The integral is determined and evaluated at the limits of integration.

The area of this region is 16.

Note: The region has a triangular shape with base (along the x-axis) of 4 units and height of 8 units. The area of a triangle is determined by the formula $A = (1/2)bh$. Since b = base = 4 and h = height = 8 for this triangle, Area = $(1/2)(4)(8) = (1/2)(32) = 16$. The area found through the triangle formula is the *same* as that found through integration.

b) A drawing of the region in question is *very* necessary.

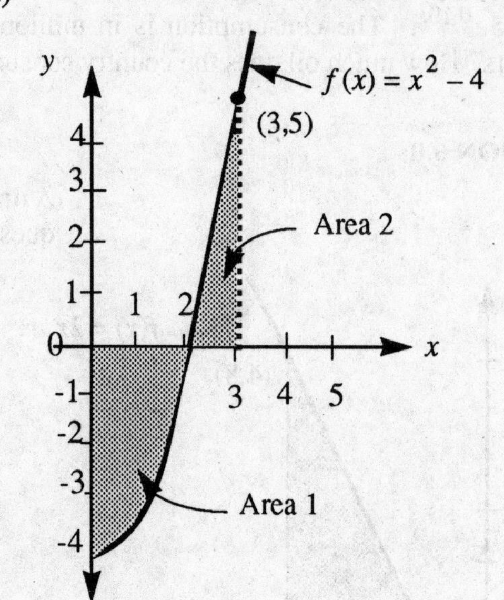

$$A_1 = \int_0^2 (x^2 - 4)\,dx$$

$$A_2 = \int_2^3 (x^2 - 4)\,dx$$

There are actually two regions in this problem. The region below the x-axis and the one above. We evaluate the *same* integrand between *two sets* of limits.

$$A_1 = \frac{x^3}{3} - 4x\Big|_0^2$$

The area of region one, A_1, is *negative*.

$$= \left(\frac{2^3}{3} - 4(2)\right) - \left(\frac{0}{3} - 4(0)\right)$$

$$= \left(\frac{8}{3} - 8\right) - (0) = \frac{8}{3} - \frac{24}{3} = -\frac{16}{3}$$

$$A_2 = \left.\frac{x^3}{3} - 4x\right|_2^3$$

The area of region two, A_2, is *positive.*

$$= \left(\frac{3^3}{3} - 4(3)\right) - \left(\frac{2^3}{3} - 4(2)\right)$$

$$= \left(\frac{27}{3} - 12\right) - \left(-\frac{16}{3}\right) = (9 - 12) + \frac{16}{3}$$

$$= -3 + \frac{16}{3} = -\frac{9}{3} + \frac{16}{3} = \frac{7}{3}$$

$$A_1 = \frac{16}{3} \left(\text{not } -\frac{16}{3}\right)$$

An area cannot, in a physical sense, be negative. Therefore, we assign A_1 a *positive value.*

$$\text{Total Area} = A_1 + A_2 = \frac{16}{3} + \frac{7}{3}$$

The total area is found by adding the two regions together.

Note: Any area occurring *below* the x-axis will be *negative* if the lower limit is less than the upper limit. All we need to do is write the area as a positive number.

Note: The problem was split into two areas, one below the x-axis and one above the x-axis. The graph allowed us to see that the function, $x^2 - 4$, crossed the x-axis at 2. The number, 2, then became the upper limit for A_1 and the lower limit for A_2. If we had integrated from 0 to 3 directly we would have gotten an erroneous value of 3 (actually –3) for the total area. In effect, the lower area would have subtracted the upper area out. That is why a *graph is very important.* We would not have known to split the problem into two integrals or to add areas if the graph were not present.

c) $C'(t) = 15e^{0.10t}$

The consumption rate is given.

The graph of the function is below. The *area, A,* is the *total consumption* for 10 years.

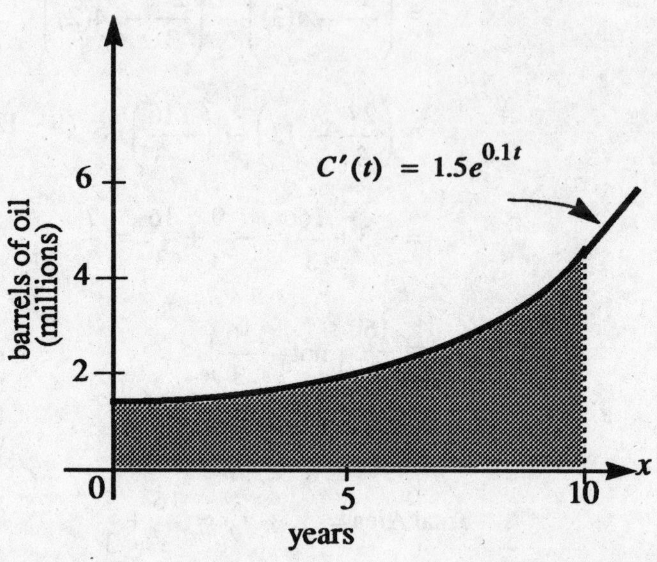

$$\text{consumption} = \int_0^{10} 1.5e^{0.10t}\,dt$$

The integral is established.

let $u = 0.10t$

$$\frac{du}{dt} = 0.10$$

$$du = 0.10dt$$

We let u equal the exponent. du must equal $0.01t$.

$$1.5\int_0^{10} e^{0.10t}\,dt$$

We need 0.10 in front of dt. Place 0.10 next to dt and put 10 outside the $\int$ symbol. $(0.10 \times 10 = 1)$

$$1.5\,(10)\int_0^{10} e^{\overbrace{0.10t}^{u}}\,\underbrace{0.10dt}_{du}$$

$$15\int e^u du$$

$$= 15e^u + C$$

The pattern follows Rule II of Table 6.5. (The limits are temporarily not shown.)

$$= 15e^{0.10t} + C$$

The integral is found and $u = 0.10t$ is replaced.

$$= 15e^{0.10t}\Big|_0^{10}$$

The limits are returned.

$$= 15e^{0.10\,(10)} - 15e^{0.10\,(0)}$$

The definite integral is evaluated.

$$= 15e^1 - 15e^0$$

$$= 15e - 15\,(1) = 15e - 15$$

$$\cong 15\,(2.718) - 15 \cong 25.77$$

The result is the total oil consumption for 10 years (in millions of barrels).

Total consumption = 25.77 million barrels of oil.

Consumer Surplus

We are delighted to find a bargain. When we pay less than we expect to pay for an item, the difference is called a *consumer surplus*.

The formula for consumer surplus is:

6.4 Consumer Surplus Formula

$$\text{Consumer Surplus} = \int_0^{x_0} (p\,(x) - p_0)\,dx$$

where:

x_0 is the amount of the items produced at the equilibrium price
$p(x)$ is the demand function
p_0 is the equilibrium price

EXAMPLE 6.9

What is the consumer surplus for the demand function:

$p(x) = 200 - 10x - x^2$

The equilibrium price occurs at the production of 5 units per day.

SOLUTION 6.9

$p(x) = 200 - 10x - x^2$ — Copy the demand function.

$$p(5) = 200 - 10(5) - 5^2$$
$$= 200 - 50 - 25$$
$$= 125 = p_0$$

The equilibrium price is found by substituting $x = 5$ into the demand function.

$$\text{consumer surplus} = \int_0^{x_0} (p(x) - p_0)\,dx$$

We can use Formula 6.4 to set up the integral.

$$= \int_0^5 (200 - 10x - x^2 - 125)\,dx$$
$$= \int_0^5 (75 - 10x - x^2)\,dx$$
$$= 75x - \frac{10x^2}{2} - \frac{x^3}{3}\Bigg|_0^5$$

The integral is determined.

$$= 75x - 5x^2 - \frac{x^3}{3}\Bigg|_0^5$$
$$= \left(75(5) - 5(5)^2 - \frac{(5)^3}{3}\right) - \left(75(0) - 5(0)^2 - \frac{0^3}{3}\right)$$

The limits are substituted.

$$= \left(375 - 125 - \frac{125}{3}\right) - 0$$

The substitution is simplified.

$$= (250 - 41.67)$$
$$= \$208.33 = \text{consumer surplus}$$

The consumer surplus is determined.

Practice Exercises

1. Find the following sums.

a) $\sum_{n=1}^{5} n+1$

b) $\sum_{n=2}^{8} 5$

c) $\sum_{k=1}^{4} k^2+1$

d) $\sum_{j=5}^{8} 3x^j$

e) $\sum_{n=1}^{3} (-1)^n (n+2)$

2. Find the integral.

a) $f(x) = x^2+7x-2$

b) $f(x) = 4x^{1/2}+5x^{3/2}$

c) $f(r) = \dfrac{r^2+8r}{r}$

d) $f(x) = \dfrac{x^5}{\sqrt[3]{x}}$

3. Find the integral and the constant of integration for the following functions.

a) $f(x) = \dfrac{6}{x} \quad F(1) = -7$

b) $f(x) = x^2-3x+1 \quad F(0) = 9$

4. Evaluate the following definite integrals.

a) $\int_1^5 3x dx$

b) $\int_{-2}^{2} (x^2-x)\,dx$

c) $\int_1^3 5e^x dx$

5. Find the following integrals through substitution.

a) $\int 2x(x^2+1)\,dx$

b) $\int 3x^2\sqrt{x^3+2dx}$

c) $\int \dfrac{3x^2+7}{x^3+7x+2}dx$

d) $\int \dfrac{6}{x+3}dx$

e) $\int 7xe^{x^2}dx$

6. Use integration by parts to solve:

$\int (x^2+1)\,e^x dx$

7. a) Find the area bounded by $4-x^2$ and the x-axis.

b) Find the area bounded by $x = 3$, $x = 5$, $y = x^3$ and the x-axis.

Answers

1. a) 20

   b) 35

   c) 34

   d) $3x^8 + 3x^7 + 3x^6 + 3x^5$

   e) –4

2. a) $F(x) = \frac{x^3}{3} + \frac{7x^2}{2} - 2x + C$

   b) $F(x) = \frac{8x^{3/2}}{3} + 2x^{5/2} + C$

   c) $F(r) = \frac{r^2}{2} + 8r + C$

   d) $F(x) = \frac{3x^{17/3}}{17} + C$

3. a) $F(x) = 6\ln|x| - 7$

   b) $F(x) = \frac{x^3}{3} - \frac{3x^2}{2} + x + 9$

4. a) 36

   b) 16/3

   c) $5(e^3 - e)$

5. a) $\frac{(x^2+1)^2}{2} + C$

   b) $\frac{2\sqrt{(x^3+2)^3}}{3} + C$

   c) $\ln|x^3 + 7x + 2| + C$

   d) $6\ln|x+3| + C$

   e) $\frac{7}{2}e^{x^2} + C$

6. $(x^2+1)e^x - 2xe^x + 2e^x + C$

7. a) 32/3

   b) 136

Index

Index

M

N

O

P

Q

R

S

T

V

X

Y

Z

Notes/Calculations

Notes/Calculations

Notes/Calculations

Notes/Calculations

Notes/Calculations

Notes/Calculations